DISCONTINUITIES IN ECOSYSTEMS AND OTHER COMPLEX SYSTEMS

Complexity in Ecological Systems Series

Timothy F. H. Allen and David W. Roberts, Editors
Robert V. O'Neill, Adviser
Robert Rosen
Life Itself: A Comprehensive Inquiry Into the Nature, Origin, and Fabrication of Life

Timothy F. H. Allen and Thomas W. Hoekstra
Toward a Unified Ecology

Robert E. Ulanowicz
Ecology, the Ascendent Perspective

John Hof and Michael Bevers
Spatial Optimization for Managed Ecosystems

David L. Peterson and V. Thomas Parker, Editors
Ecological Scale: Theory and Applications

Robert Rosen
Essays on Life Itself

Robert H. Gardner, W. Michael Kemp, Victor S. Kennedy, and
John E. Petersen, Editors
Scaling Relations in Experimental Ecology

S. R. Kerr and L. M. Dickie
The Biomass Spectrum: A Predator-Prey Theory of Aquatic Production

John Hof and Michael Bevers
Spatial Optimization in Ecological Applications

Spencer Apollonio
Hierarchical Perspectives on Marine Complexities: Searching for Systems in the Gulf of Maine

T. F. H. Allen, Joseph A. Tainter, and Thomas W. Hoekstra
Supply-Side Sustainability

Jon Norberg and Graeme S. Cumming, Editors
Complexity Theory for a Sustainable Future

David Waltner-Toews, James J. Kay, and Nina-Marie E. Lister, Editors
The Ecosystem Approach: Complexity, Uncertainty, and Managing for Sustainability

DISCONTINUITIES IN ECOSYSTEMS AND OTHER COMPLEX SYSTEMS

Craig R. Allen and
C. S. Holling, Editors

COLUMBIA UNIVERSITY PRESS NEW YORK

COLUMBIA UNIVERSITY PRESS
Publishers Since 1893
New York   Chichester, West Sussex

Library of Congress Cataloging-in-Publication Data

Discontinuities in ecosystems and other complex systems / Craig R. Allen and C. S. Holling, editors.
p. cm.— (Complexity in ecological systems series)
Includes bibliographical references.
ISBN 978-0-231-14444-5 (cloth : alk. paper)—ISBN 978-0-231-14445-2 (pbk. : alk. paper)
—ISBN 978-0-231-51682-2 (ebook)
1. Ecology–Statistical methods. 2. Discontinuous groups. I. Allen, Craig R. II. Holling, C. S. III. Title. IV. Series.
QH541.15.S72D57 2008
577—dc22 2008006382

Columbia University Press books are printed on permanent and durable acid-free paper.

This book is printed on paper with recycled content.
Printed in the United States of America

c 10 9 8 7 6 5 4 3 2 1
p 10 9 8 7 6 5 4 3 2 1

References to Internet Web sites (URLs) were accurate at the time of writing. Neither the author nor Columbia University Press is responsible for URLs that may have expired or changed since the manuscript was prepared.

CONTENTS

PREFACE

THE SCALING of physical, biological, ecological, and social phenomena has become a major focus of efforts to develop simple representations of complex systems (Bak 1996; West, Brown, and Enquist 1997; West and Brown 2005). From those representations have come the identification, explanation, and testing of scaling laws and the classification of appropriate scales of analysis. But there has been little focus on the significance of departures from those scaling relationships or of departures from unimodal continuous distributions. Some of those departures mark breaks in dominant scaling relationships, and they separate different scaling regimes. Other departures mark concentrations of data along them.

Much attention has focused on discovering universal scaling laws that emerge from simple physical and geometric processes. This work, however, is motivated by the discovery of regular patterns of departures from these scaling laws and from continuous distributions of systems' attributes (Holling 1992; Bessey 2002). The departures seem to demonstrate how living systems of animals, plants, and people develop self-organized interactions with physical processes over narrower ranges of scale. Just as pulses in time of resource acquisition by animals increase the efficiency of energy utilization, perhaps these aggregations in the morphological, geometric, and behavioral variables of animals, plants, and people have unique self-organizing properties that affect evolutionary change and thus may have potential policy consequences.

Part of this clumping structure comes from the way living systems form hierarchies. Simon (1974) argued for the adaptive significance of such structures. He called them hierarchies, but not in the sense of top-down sequential and authoritative control. Rather, semiautonomous levels are formed from the interactions among sets of variables that share similar speeds (and, we would add, spatial and morphological attributes). Each level communicates a small set of information or quantity of material to the next higher (slower and coarser) level. An example for a forested landscape is shown as figure P.1.

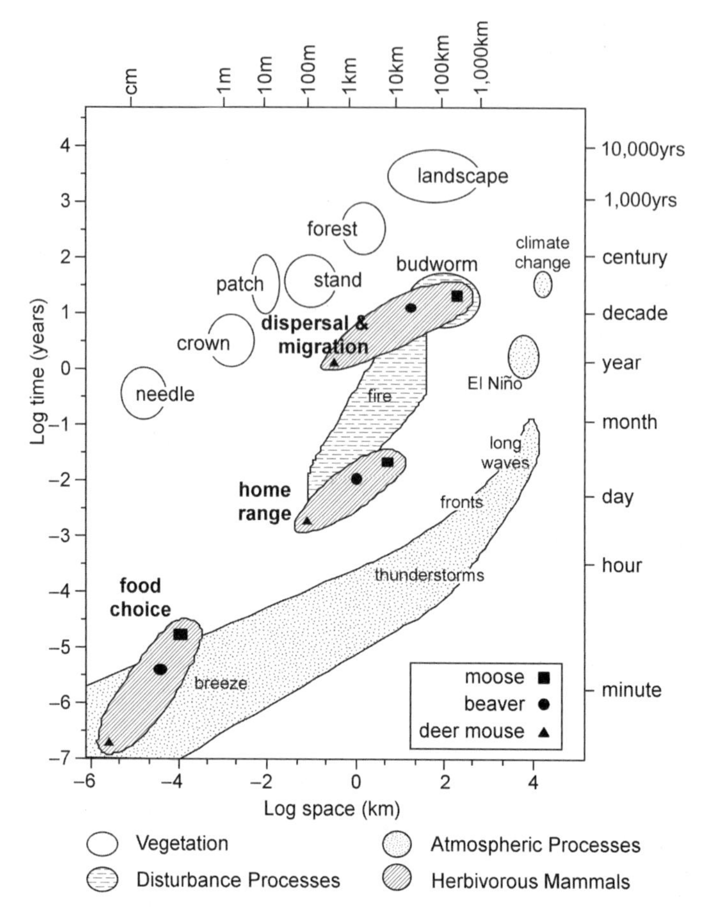

FIGURE P.1 Time and space scales of the boreal forest (Holling 1986) and their relationship to some of the processes that structure the forest. These processes include insect outbreaks, fire, atmospheric processes, and the rapid carbon dioxide increase in modern times (Clark 1985). Contagious mesoscale disturbance processes provide a linkage between macroscale atmospheric processes and microscale landscape processes. Scales at which deer mouse, beaver, and moose choose food items, occupy a home range, and disperse to locate suitable home ranges vary with their body size (Holling 1992; Peterson, Allen, and Holling 1998).

So long as the transfer from one level to the other is maintained, the interactions within the levels themselves can be transformed, or the variables can be changed without the system as a whole losing its integrity. As a consequence, this structure allows wide latitude for "experimentation" within levels, thereby greatly increasing the speed of evolution. Such structure is a characteristic of what have been termed *complex systems*.

Ecologists were inspired by Simon's seminal article to transfer the term *hierarchy* to ecological systems and to develop its significance for a variety of ecological relationships and structures. In particular, T. Allen and Starr (1982) and O'Neill and colleagues (1986) launched an expansion of theoretical understanding by shifting attention from the small-scale view that characterized much of biological ecology to a multiscale and landscape view that recognizes that biotic and abiotic processes can develop mutually reinforcing relationships.

Ecological, biological, and anthropogenic hierarchies appear to exhibit multiple scale regimes—there are breaks between levels as processes controlling structure shift from one set to another. For example, the analysis of vegetation pattern on landscapes has shown that different scaling regimes exist, each with its own fractal dimension (Krummel et al. 1987). The analyses of animal community patterns also have revealed a cross-scale pattern of multiple scale regimes, shown by the clumping of animal attributes such as size (Holling 1992). Walker, Kinsig, and Langridge (1999) have shown that in arid Australia, where they conducted their study, plants' morphological attributes form distinct aggregations that correspond to the plants' functions.

Once a discontinuous pattern of "clumps" separated by "gaps" is formed in a distribution, it closely interacts with complex sets of related and scale-specific variables. The consequences of scale-specific structure and discontinuities determine, in part, how resilient the systems are and how robust to modification by policy or by exogenous change. For example, understanding the scaled nature of animal communities and the scale breaks intrinsic within them has led to a better understanding of the manner in which ecological resilience is generated from the distribution of biological diversity. There are two types of such diversity: a diversity that affects biological function within a range of scale (Walker, Kinsig, and Langridge 1999) and a diversity that affects biological function across scales (Peterson, Allen, and Holling 1998). Some have suggested that biological diversity serves a purpose similar to that of redundant design features that an engineer might use to achieve engineering reliability. But the within-scale and cross-scale types of diversity we have identified cannot be satisfactorily described as simple redundancy. Rather, the species in a suite of species interacting within the same range of scale (i.e., at the same level in a hierarchy; species of similar size) have similar but not at all identical effects and functions. They differ in the particular function and in their degree of influence and sensitivity to change. This kind of diversity produces an overlapping reinforcement of function that is remarkably robust. We call it *imbrication*. Other seemingly redundant species, those species that seem to perform the same function (e.g., seed dispersal), are in reality not redundant because they perform those functions at distinctly different ranges of scale in terms of both space and time. We call this relationship *cross-scale reinforcement*. Interestingly, recent work has hinted at similar relationships in other complex systems.

The richness of function within aggregations of firms of similar size is related to resilience in that employment volatility is lower with increasing richness of firm functions (Garmestani et al. 2006).

The discontinuities that characteristically separate clumps or aggregations in attributes of animal communities, such as body sizes, correlate strongly with a set of poorly understood biological phenomena that seem to mix contrasting attributes. These phenomena include invasion, extinction (high species turnover), high population variability, migration, and nomadism. For example, Allen, Forys, and Holling (1999) have shown that the body masses of endangered and invasive species in a community occur at the edges of body-mass clumps (i.e., at scale breaks) two to four times as often as expected by chance. That correlation is consistent in all eight data sets so far examined. Those sets include four different taxonomic groups (birds, mammals, herpetofauna, and bats) in two different ecosystems (Mediterranean climate and wet savannas). The association of declining species with discontinuities is confirmed in Skillen and Maurer's analysis in chapter 11 of this volume. The clustering of these phenomena at predictable locations near the discontinuities that mark scale breaks suggests that there is variability in resource distribution or that availability is greatest at those breaks. Is this true in other systems?

There has been some skepticism regarding whether discontinuous, clumpy distributions are the norm and are real. Part of that skepticism is because some apparent patterns in nature proposed in the past have subsequently been shown to be artifacts. The ecological and biological literature has historically been dominated by assumptions that organisms' attributes are distributed continuously, not discontinuously, and that such distributions are unimodal and continuous. Manly (1996) applied a conservative statistical test to the original data sets presented by Holling (1992) and concluded that, at most, two clumps or aggregations of body mass were significantly present, rather than the eight or more that Holling identified. Conservative tests, of course, reduce the chance of being wrong (Type I error), but they also reduce the chance of being able to detect a real pattern (Type II error).

The nature of discontinuities in data from ecological, economic, and social systems remains relatively unexplored. Careful investigation of the partitioning of diversity within and across scales, of the importance of clumps and discontinuities in generating resilience, and of the phenomena nonrandomly associated with discontinuous structure in complex systems may lead to fruitful avenues of investigation in the analysis of ecological, economic, and social systems. For example, high (resource) variability can be both a detriment (i.e., the association between extinct and declining species and discontinuities) or a boon (the success of invasive species and development of nomadic behavior at discontinuities). Under what circumstances is it beneficial to exploit or specialize in discontinuities between distinct ranges of scale? When a system's resilience is exceeded (or the limit of its resilience is approached), do species or organizations associated with discontinuities and the edges of clumps unravel first? What economic or social indicators exhibit discontinuities or clumps in data?

The scaling of the distribution of city sizes and the sizes of economic fluctuations have been analyzed, but principally as a search for universal scaling laws. Do those distributions also demonstrate such clumpy departures from the scaling laws? And if so, what is the cause and consequence of the distributions? What other relevant social and economic variables exhibit scale breaks that define aggregated, discontinuous distributions of attributes? What do these behaviors mean? How does the segregation of systems into discreet scales of influence increase systems' resilience?

Beginning in February 1999 in Atlanta, Georgia, and continuing until at least December 2007 in Santiago, Chile, we have held small meetings of ecologists, social scientists, economists, mathematicians, and statisticians to discuss these ideas and their relevance to resilience and change in dynamic systems. The goal of the workshops has been to share information across the disciplines of ecology, economics, and social sciences. We offer some of our discoveries in this volume. Many of the chapters herein were originally spawned from a meeting at the Santa Fe Institute in 2001; that meeting specifically brought together both proponents and skeptics to discuss these ideas in the unique environment that is the Santa Fe Institute. Our focus is specifically on the discontinuities that separate distinct ranges of scale in complex systems. The location of discontinuities can be detected with a number of statistical approaches (see chapters 9 and 11, for example), and the variables in complex systems, such as species in ecosystems, that are proximate to discontinuities appear to exhibit the high variability that is theoretically predicted for scale breaks in biotic systems (Wiens 1989). Discontinuity theory focuses not so much on within-scale structure in complex systems, but on the transitions between scales and the phenomena associated with those transitions.

The twelve chapters herein are organized into three parts. The first part focuses on background material and some contrasting views on how complex systems discontinuously organize. The second part focuses on discontinuous patterns noted in a number of different systems and on the detection of those patterns. The third part touches on the potential significance of discontinuities in complex systems. Note that we have attempted to be consistent in our use of terminology throughout this volume, but different authors have used different terms to describe discontinuous distributions over time. Discontinuities have been described as *discontinuities* or as *gaps*. The groupings of variables defined by discontinuities have been variously described as *lumps*, *clumps*, and *aggregations*. Some authors in some circumstances have focused on multimodality rather than on discontinuity per se. We use the terms *clump* and *aggregation* to designate the grouping of species or other variables that are bounded by discontinuities, or gaps.

The analysis of discontinuities in complex systems is in its infancy. A focus on power laws still dominates much of science, including ecology and urban studies, to the point that power laws have been trivially fit to almost everything (e.g., Kirchner and Weil 1998). We are convinced that such approaches mask the interesting dynamics of systems and that those dynamics are best revealed by investigating deviations from assumed continuous or power law distributions.

DISCONTINUITIES IN ECOSYSTEMS AND OTHER COMPLEX SYSTEMS

PART ONE

BACKGROUND

1

PANARCHIES AND DISCONTINUITIES

Crawford S. Holling, Garry D. Peterson, and Craig R. Allen

WE DESCRIBE the organization of ecological systems as *panarchies* (Gunderson and Holling 2002). Panarchies are hierarchically arranged, mutually reinforcing sets of processes that operate at different spatial and temporal scales, with all levels subject to an adaptive cycle of collapse and renewal, and with levels separated by discontinuities in key variables. Dominant processes entrain other processes to their spatial and temporal frequencies. This entrainment (an interaction among process and structure dominated by one or a few processes within a range of scale) produces discontinuities (gaps) and aggregations (clumps) in animal community body-mass distributions and in distributions of other complex systems, such as urban systems (Garmestani, Allen, and Bessey 2005; chap. 8 in this volume) and economic systems (chap. 10 in this volume). We provide some examples of discontinuous and clumpy distributions and discuss their potential consequences for ecological theory.

The original suggestion that ecosystem attributes might be distributed discontinuously (or in a clumpy way) came from a review of twenty-seven different examples of ecosystems subject to different types of management (Holling 1986). The essential ecological dynamics of these systems can be described with three sets of variables, each set operating at a qualitatively different speed. In the forests of eastern North America, for example, those variables are: with a one-year generation time, an insect—the spruce budworm—and conifer needles; the tree crowns with a twelve- to fifteen-year generation; and the trees with a one-hundred-year-plus generation time. Similar sets of variables were identified in other insect forest systems, rangelands, and fisheries and within plant and human disease systems, all of which therefore have a discontinuous pattern to their time dynamics. These discontinuities are also reflected in the spatial patterns visible in these systems. The fast variables are small, and the slow variables large. Hence, the three critical sets of variables can be drawn from a larger hierarchy of the sort suggested in figure 1.1. The key point is that each element in this hierarchy is functioning in time and space at its own rate

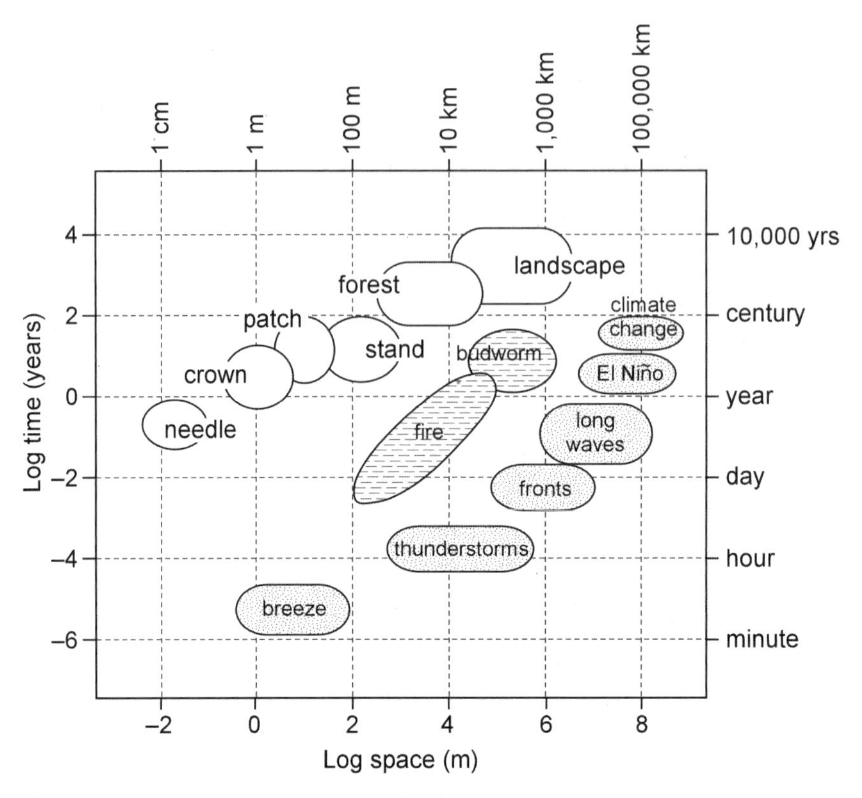

FIGURE 1.1 Time and space scales of the boreal forest and of the atmosphere and the relationship of these scales to some of the processes that structure the boreal forest. Contagious mesoscale processes such as insect outbreaks and fire mediate the interaction between faster atmospheric processes and slower vegetation processes. (Reprinted from Gunderson and Holling 2002 with permission of Island Press.)

and that these rates jump from one level of the biotic components of this hierarchy to another.

Moreover, in these systems, the dynamics of each set of the components within a range of scale follow an adaptive cycle of change. The adaptive cycle model proposes that the internal dynamics of systems cycle through four phases: rapid growth, conservation, collapse, and reorganization. As unorganized processes interact, some processes reinforce one another, rapidly building structure and organization. This organization channels and constrains interactions within the system. However, the system becomes dependent on structure and constraint for its persistence, leaving it vulnerable to either internal fluctuations or external disruption. The system eventually collapses, allowing the remaining disorga-

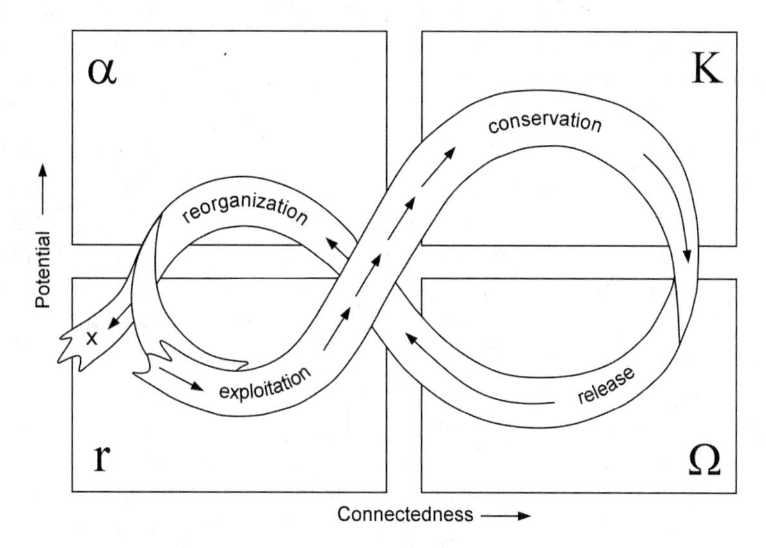

FIGURE 1.2 A stylized representation of the four ecosystem functions *(r, K, Ω, α)* and the flow of events among them. The arrows show the speed of that flow in the cycle, where short, closely spaced arrows indicate a slowly changing situation and long arrows indicate a rapidly changing situation. The cycle reflects changes in two properties: (1) *y* axis, the potential inherent in the accumulated resources of biomass and nutrients; and (2) *x* axis, the degree of connectedness among controlling variables. The exit from the cycle indicated at the left of the figure suggests, in a stylized way, the stage where the potential can leak away and where a flip into a less productive and organized system is most likely (Holling 1986). (Reprinted from Gunderson and Holling 2002 with permission of Island Press.)

nized structures and processes to reorganize or the structures and processes from neighboring, extrinsic systems to invade. This cycle is summarized in figure 1.2. These two features—the hierarchy and its nested adaptive cycles—define a dynamic, discontinuous template that entrains other variables. They define a panarchy (Gunderson and Holling 2002).

Panarchy is the term devised (Gunderson and Holling 2002) for a concept that helps explain the evolving nature of complex adaptive systems. A panarchy is the hierarchical structure in which systems of nature (e.g., forests, grasslands, lakes, rivers, and seas), of humans (e.g., governments, settlements, businesses, and cultures), and of combined human-nature systems (e.g., agencies that control natural-resource use) are interlinked via adaptive cycles of growth, accumulation, restructuring, and renewal. These transformational cycles take place in nested sets at scales ranging from a leaf to the biosphere over periods from days to geologic epochs and from a family to a sociopolitical region over periods from years to many centuries. By understanding these cycles and their scales, it may be possible to evaluate their contribution to sustainability, to identify the points at which a system is capable of positive change, and to indicate the points where it is vulnerable. These leverage points may theoretically be used to foster resilience and sustainability within a system.

The concept of the adaptive cycle and the observation that key variables operate at distinct and separate scales emerged from a synthesis of empirical studies (Holling 1986); it was not deduced from first principles. Theory did not determine what observations were made; rather, observations of nature and the practice of ecological management dictated theory. Is panarchy, however, a consequence of the way analysts and modelers make convenient modeling decisions and simplifications, or is it an accurate depiction of the way ecosystems, industry, and management actually organize and function?

It helps that the regional models (Holling 1986) were based on extensive knowledge and analysis of actual ecological processes and that the parameters were for the most part independently estimated in the field. Predictions of some of the critically informing studies, such as the budworm/forest system (W. Clark, Jones, and Holling 1979; Holling 1986), were extensively tested by comparing them to observed behavior from different regions of eastern North America that have radically different climatic conditions and forest dynamics. The models consistently had strong predictive powers even in such extreme, limiting conditions.

Viewing nature as a panarchy has been useful because it has suggested a number of relevant hypotheses. Searching for cycles of collapse and reorganization in ecosystems has suggested new perspectives in the understanding of ecological dynamics. Searching for minimal sets of critical structuring variables that are separated in scale—both speed and size—has led to useful simplifications of ecological dynamics in example after example (Gunderson and Holling 2002). Furthermore, the possibility that nature is organized as a panarchy suggests that key elements of living systems, social as well as ecological, self-organize to provide a discontinuous template in space and time that provides discontinuous opportunities for other variables. This suggestion has generated a set of hypotheses to examine the discontinuous structure of a diverse set of systems.

DISCONTINUITIES

There are two reasons why an ecological panarchy might create discontinuities in the organization of a system. First, the discontinuous nature of the processes that form elements of the panarchy would create a disjunct separation of scales among key structuring variables, such as the three distinct speeds of the key variables in boreal ecosystems mentioned earlier. Second, according to the nature of the adaptive cycle itself, the four phases of the cycle are distinct, and the shift in controls from one phase to another is abrupt because the processes controlling the shift are nonlinear and the behavior multistable. Each phase creates its own distinct conditions that in turn define distinct size and speed attributes of the aggregates that control the phase or are adapted to its conditions.

K-selected species and firms tend to be big and slow; *r*-species and firms tend to be small and fast. We are not suggesting that the four phases of a cycle entrain four clumps. Rather, we suggest that the combination of panarchy-level discontinuities and adaptive-cycle discontinuities will generate a number of aggregations, the number defined by the resolution of the observations and the range of scales tested. We argue that panarchies form a discontinuous template that entrains morphological and behavioral attributes of organisms and other variables in complex systems.

By *clumps*, we mean the discrete aggregates that Krugman (1996) describes and explains for human settlements—cities, towns, and villages. He isolates centripetal and centrifugal forces that produce instabilities that generate agglomerative patterns and discrete aggregates. Although there are discrete aggregates in ecosystems such as individual organisms, most aggregates are amorphous, such as plant associations, ecosystems, and even, some would argue, species. We also propose that the attributes of these discrete aggregates are themselves distributed in an aggregated manner. These attributes include the periodicities of fluctuations, the size of objects at different scales on a landscape, the scales of animals and humans' decision processes, and animals and plants' morphological and functional attributes.

The distributions of attributes, our proposition states, will not be continuous or unimodal. Rather, they will be discontinuous (with gaps in a distribution) or multimodal or both. Similarly, scaling relations will produce clusters of attributes along regression lines (clumps) or patterns in residuals that indicate breaks between scaling regimes.

In contrast to this proposition, much of modern science, including ecology, seeks simplifying, universal laws by searching for continuous, unimodal properties. For example, the scaling of physical, biological, ecological, and social phenomena has become a major focus of efforts to develop simple and universal representations of complex systems (Gell-Mann 1994). This focus has led to the identification, explanation, and testing of scaling laws for systems as wide ranging as biophysical systems (Bak 1996; West, Brown, and Enquist 1999), ecological systems (Keitt and Stanley 1998; Enquist et al. 2003), firms and countries (Brock and Evans 1986; Stanley et al. 1996), and humans (Krugman 1996). However, as stated in the preface, there has been little focus on the pattern and dynamics of departures from those scaling relationships—either as clustering of attributes (clumps) or as breaks (discontinuities or gaps) between two scaling regimes. The application and description of scaling laws has clearly been useful, but it is not without drawbacks. The most interesting dynamics of a system can be overlooked by painting them with too broad a brush and ignoring both the deviations from scaling laws as well as the causes and consequences of dynamics and structures found at those deviations. Brock (1999) reviews and discusses the perils and pitfalls of the application and interpretation of scaling laws in economics.

EVIDENCE FOR DISCONTINUITIES

Holling (1992) observed discontinuous, clumpy distributions of bird and mammal body masses in a variety of ecosystems. Evidence is accumulating that body masses are distributed in a discontinuous manner both on land and in water, whose cause must be associated with slow, conservative properties of landscapes and waterscapes, and with mutually reinforcing or self-organizing relationships.

Attributes other than animal body mass also demonstrate discontinuities. For example, Walker, Kinsig, and Langridge (1999) show that plants' morphological attributes have aggregated distributions and that each clump corresponds to a functional role of plants in an ecosystem. They demonstrate that functionally significant morphological attributes of grass and forb species show three to five clump clusters in savanna ecosystems. Garmestani, Allen, and Bessey (2005; see also chap. 8 in this volume) and Bessey (2002) demonstrate deviations from power laws and discontinuous distributions in regional distributions of city size. Garmestani and colleagues (2006) further demonstrate discontinuities in distributions of firm size and a relationship between richness of firm function within aggregations and employment volatility within an industrial sector.

There has been skepticism that clumps are real, and some ecologists and statisticians have applied their own tests to determine if discontinuities exist in animal body-mass distributions. But they either have applied extremely conservative methods biased to detecting unimodal distributions (Manly 1996) or have asked a different question than the one relevant for testing the proposition discussed here (Siemann and Brown 1999; see also chap. 9 in this volume). Siemann and Brown's test concerned the sizes of individual gaps, not the existence of a pattern of aggregations and gaps within a data set.

More convincing tests come from proposing and invalidating alternative hypotheses of causation. It is those tests, together with appropriate statistical ones of the kind suggested by Manly (1996), that can lead to multiple lines of evidence that converge on a credible line of argument. It took more than three decades to confirm the existence and management significance of multistable states in ecosystems (Carpenter 2000). It might take as long to establish the reality, cause, and significance of discontinuities and aggregations. Much of the controversy regarding the prevalence of discontinuities within ecosystems and other complex systems has recently diminished, however. That discontinuities are real has come to be accepted, but it is not universally agreed upon what mechanisms produce them. Craig Allen and colleagues (2006) provide a thorough discussion and weighting of evidence that supports or refutes different mechanistic hypotheses underlying discontinuous animal body-mass distributions. We briefly discuss a subset of these hypotheses in the next section.

CAUSES OF DISCONTINUITIES

At least three proximate causal mechanisms can directly produce discontinuous distributions of body masses: the cross-scale ecological template produced by a panarchy, gaps due to limited representation of possible body masses within a local community, and species interactions within ecological communities. Together they represent both slow and fast processes.

ECOLOGICAL TEMPLATE

Panarchies can produce landscapes that are composed of different-size resource aggregations at different scales. Each size of resource aggregation reflects the influence of one or a few dominant ecosystem processes. There are well-established allometric relationships between an animal's body size and its energy needs, speed, distance of movement, and life span (e.g., Peters 1983). As a consequence, not all sizes of animals can survive in a given landscape—only those whose scaled physiological, behavioral, and life-cycle features match the scale of resource availability and those whose morphology allows them to forage effectively and otherwise interact with a specific scale of structure within a landscape. Morton (1990) used that possibility to explain the extinction of middle-size mammals after European settlement in the arid interior of Australia. He proposed that altered fire regimes, the impacts of introduced rabbits on vegetation, and predation by introduced foxes reduced the resources in patches at intermediate scales and increased mortality of mammals with body sizes adapted to exploiting resources at those mesoscales.

There is empirical evidence that biological and ecological attributes of specific landscapes exhibit multiple scaling regimes, that there are breaks between scales as processes controlling structure shift from one set to another, and that there is clustering of attributes at distinct scales. These shifts are suggested in figure 1.1, but formal analysis of vegetation pattern on landscapes has shown that different scaling regimes exist, each with a distinct fractal dimension (Krummel et al. 1987).

Furthermore, analyses of animal communities on specific landscapes have revealed a cross-scale and multimodal—or discontinuous—pattern in animal body-mass distributions (Holling 1992). Architecturally simple landscapes have few clumps in the body-mass distributions of animals living in them; complex landscapes have many clumps. For example, Schwinghamer (1981) and Raffaelli and colleagues (2000) show that architecturally simple marine sediments have communities living within them that have three and perhaps four clumps in their inhabitants' size distributions. Boreal forest landscapes (Holling 1992) are somewhat more complex; their mammal and bird communities show about eight clumps in body mass. Tropical forests systems are even more complex, and their bird inhabitants show an even greater number of clumps (Restrepo, Renjifo, and Marples 1997). Consequently, we expect that an increase in habitat structural complexity will also increase the complexity of lump structure because that "structural complexity" is a reflection of a richer array of distinct scales of structure and resource aggregation and distribution.

LIMITED REPRESENTATION

Phylogenetic constraints also reflect the operation of slow processes that might explain clumps because organisms might have evolved a limited number of body sizes that can function efficiently. That is, evolution may produce a clumpy universe of species from which communities are assembled. Even an otherwise randomly assembled community will have a discontinuous body-mass distribution because its members are drawn from a species pool that is also discontinuous (a narcissus effect; Moulton and Lockwood 1992) due to inherent evolutionary constraints or optimization regardless of the landscape template. Alternatively, a nonrandom set of species might have colonized an area due to reasons other than the scales of opportunity available within a given landscape, and their size distribution subsequently constrained their descendants' sizes (Marquet and Cofré 1999).

COMMUNITY INTERACTIONS

Competitive and trophic relationships are faster processes that can also produce clumps. Roughgarden (1998), for example, shows that aggregated distributions can be produced in a simple model that combines limits to species' population growth with size-dependent competition; Scheffer and van Nes (2006) reach a similar conclusion. This model produces discontinuous distributions through dynamics similar to Krugman's (1996) model of spatial aggregation. Trophic relationships also can produce clumpy distributions of body size. If predation is size limited, a "big animals eating little ones" scenario reinforces the existence of species at specific body sizes, while decreasing the resources available to species at intermediate body sizes, producing a clumpy body-mass distribution (Carpenter and Kitchell 1993).

PERSISTENCE OF DISCONTINUOUS PATTERNS

Evidence to distinguish these alternatives is accumulating. Havlicek and Carpenter (2001) analyzed data on species, populations, and sizes of phytoplankton, zooplankton, and fish collected over many years from eleven lakes in Wisconsin. All lakes showed size distributions of species with an extensive clumpy structure. Moreover, that structure was very similar in all lakes, even though the lakes differ widely in area, depth, nutrient status, food web structure, species composition, and productivity. The same clump structure remained after experimental additions of phosphate and removal of fish produced substantial differences in community structure, primary production, nutrients, chlorophyll, and bacterial production. Despite large differences in the lakes' species composition, community structure, and physical/chemical characteristics, many of the aggregations and discontinuities persisted at similar size ranges across all lakes and treatments.

Raffaeli and colleagues (2000) demonstrated the same conservative nature of body-mass structure on a smaller scale. They perturbed enclosures of marine littoral sediments in a way that changed trophic structure, species composition, and sizes of communities. The body-size structure was little affected, which suggests that aggregations and discontinuities are highly conservative features, reflecting slow processes that structure panarchies across scales. Forys and Allen (2002) analyzed the difference in aggregations and discontinuities in vertebrates in the Everglades ecosystem from the precolonial period (prior to modern species declines and extinctions) to the present time (with both declining and nonnative invasive species present) and to a hypothesized future state (eliminating species that were declining earlier). The Everglades ecosystem is in a state of biotic transformation, and nearly one-third of the vertebrates in it are declining, state or federally listed, or recently extinct, and another one-third consists of now established non-native species. Despite that enormous turnover, the pattern of discontinuities and aggregations has changed little over time. The overall structure of the system appears to be highly conservative, self-organizing, and resistant to changes in the identity of individual system elements (species)—a complex adaptive system.

It takes the kind of extreme disturbances seen over paleoecological time and space scales to change the pattern of aggregations and discontinuities in body-mass structure significantly. Eleven thousand years ago, for example, all the very large herbivores, such as giant ground sloths and the shovel-tusked elephants, became extinct in North and South America in less than one thousand years (Martin 1984). Lambert and Holling (1998) analyzed two reconstructed fossil data sets from either side of the North American continent to identify the body-mass structure before and after that massive extinction pulse. The data demonstrate a significant body-mass structure that remained entirely unchanged for animals less than 41 kg, even though extinction occurred among those species. Replacement by new species of similar sizes maintained the structure despite species turnover. At greater than 41 kg, the body-mass structure was entirely transformed, and the largest clump of animals, with masses greater than 1,000 kg, was eliminated entirely. At the end of the most recent ice age, climate change, altered fire regimes, and hunting by a new, efficient hunting culture conspired to change completely the habitat template at coarse scales, but only at coarse scales.

It is plausible, moreover, that the large herbivores created and maintained that coarse pattern of grasslands and forest in the manner proposed by Zimov and colleagues (1995) for the megaherbivores of northern Russia and Alaska during the same period. Grazing by the large herbivores likely created and maintained vegetative patterns appropriate for their own existence, just as large herbivores still do in Africa (Owen-Smith 1988). They likely were part of a set of critical, systemic self-organizing processes that created a slow, large, adaptive cycle at coarse scales. Such self-organizing processes and the adaptive cycles they create are very resilient, but once they collapse, they unravel quickly through positive feedbacks. Thus, in the case from eleven thousand years ago, one slow, large level of the panarchy collapsed, explaining the sudden and continental scale of the species extinctions resulting from that transformation. The collapse did not cascade to smaller scales, however, so the body sizes appropriate to those smaller scales remained unchanged.

The conservative, persistent structure of discontinuous body-mass distributions suggests that the robust, sustaining features of the panarchy described earlier are formed by slow ecological and evolutionary processes. The distribution of aggregations and discontinuities may be used as a type of bioassay of the structure of a panarchy. Although clumps themselves are relatively stable, populations of species within them are not—they are highly labile and reflect the effect of the stochastic processes, competition, and dynamic changes that structure adaptive cycles.

SIGNIFICANCE OF DISCONTINUITIES AND AGGREGATIONS

Once the pattern of discontinuities and aggregations are formed in a distribution, they entrain a complex set of related variables. The consequences determine, in part, how resilient the patterns are and how robust to modification by policy or by exogenous change. A system's resilience is influenced by both within-scale and cross-scale diversity of function.

For example, the boreal forest's properties and patterns are maintained by a set of processes involving an insect defoliator, the spruce budworm, two species of trees, and avian predators of the budworm. The thirty-five species of insectivorous birds are critical. They are distributed across five body-mass clumps (Holling 1988). Species in the same clump are most likely to compete with one another because they forage at similar scales. But they have different responses to climatic and other environmental changes and are diverse in their ecological functions. In each size category, there is at least one insectivorous species that includes budworm in its diet.

Species in different body-mass aggregations forage at different scales, so they begin to utilize budworm only when budworms are at an appropriate level of aggregation—that is, only when the scale of the budworm resource matches the scale of the insectivores' foraging and environmental interaction. Small warblers, for example, respond to single budworms on branches, larger sparrows to aggregations of budworms on branches, and still larger grosbeaks to aggregations within crowns. Hence, as budworm population densities increase during periodic outbreaks, a strong biotic resistance develops that brings into play more and larger avian predator species with larger appetites and from more distant areas. All the bird species in question are foliage-gleaning insectivores and would be considered redundant in some models of the relationship between diversity and stability, but because each species interacts with its environment at a different scale, the different species are not really redundant, but instead provide a cross-scale reinforcement of function. When the regulation of budworms eventually fails, it does so suddenly and over large spatial scales of hundreds of kilometers, allowing the budworm to emerge as a forest-structuring process as it defoliates and kills large areas of forest.

Diversity of functional types of plants contributes to resilience and persistence of functions in a similar way, as Walker, Kinsig, and Langridge (1999) demonstrated when they compared savannas exposed to different intensities of grazing. We suppose that the variety of grazer and browser species in African savannas also provides a wide range of both within-scale and between-scale control to perturbation, thus generating system resilience.

The effect of diversity of function is not redundancy in the sense that an engineer might replicate function to achieve engineering reliability. Rather, each species in the same size aggregation has a similar scale of resource exploitation, but at least a slightly different function and different responses to unanticipated environmental change. If the ecosystem were a theater, the species within an aggregation would be the stand-in actors who are prepared to replace each other in the event of unexpected external surprises and crises. Species in different aggregations also can engage in similar or related ecosystem functions, but because of their different sizes, they differ in the scale and degree of their influence. In our ecosystem theater, species in different clumps are the actors waiting in the wings to facilitate a change in pace or plot when needed. The within-scale diversity produces an overlapping reinforcement of function that is remarkably robust, and when within-scale diversity is combined with cross-scale reinforcement, the distribution of function within and across scales adds tremendously to the resilience of ecological systems.

The same kind of imbricated redundancy is a common property of many biological phenomena. For example, body temperature in homeotherms is regulated by five different physiological mechanisms ranging from metabolic heat generation to evaporative cooling. Each operates over different ranges of temperature with different efficiencies and speed of feedback control. The result is remarkably robust regulation of temperature around a narrow range. In another behavioral example, migratory birds navigate with great success between summer and winter feeding areas over enormous distances by using at least four different signals for direction—magnetic, topographic, aural, and sidereal—each of which has different levels of precision and accuracy. It is the overlapping, reinforcing nature of those separate mechanisms that makes the total effect so robust.

DECISION PANARCHIES

If animals live in a world that is organized as a panarchy, it is reasonable to expect that their behaviors reflect that organization. Objects encountered by animals are either edible or frightful or lovable or ignorable or novel. The first three define the resources on the landscape that are needed to provide food, protection, and opportunity for survival and reproduction. The latter two are items that should simply be forgotten or should be investigated for the potential they might represent. That is, forgetting, curiosity, and memory are essential in order to develop rules that are flexible and adaptive enough so that a species can persist in a fluctuating, changing world.

All five kinds of objects are created or sustained by the template formed by the ecosystem/landscape panarchy and by external introductions, events, and variability. Because the template formed by the panarchy is remarkably conservative and persistent, animals can develop rules for actions that take advantage of that persistence while retaining enough flexibility to adjust to variability and the unexpected. That is, those decision rules have the features of the adaptive cycle—both conservative and adaptable.

The evolved behavioral rules minimize information needs and processing. The rules that persist are those that make the least demand on information and contribute the most to survival and reproduction over long periods. They are not detailed, accurate, and precise, but rather economical—just sufficient—and adaptive. And if some decisions do not encounter or generate variability, they can gradually become more and more stereotyped and automatic. A simple example is the entrained rules a person learns in driving to and from work along the same route. Among insects and birds, there are many examples of rules that become genetically encoded and guide instinctive behaviors. In humans, such rules can become encoded in the myths and rituals of culture.

Holland and colleagues (1989) and Holland (1995) describe these rules as *schemas* or *scripts* wherein information stored in clusters serves to generate plausible inferences and solutions. When unexpected events occur that provide a poor match with experience, new rules can form out of the stored bits and pieces as they become recombined in novel ways, much as described for the adaptive cycle. This process of bricolage (Lévi-Strauss 1962), combined with self-organization, is as central to the formation of rules for decision as it is for forming biological or ecological structures.

Such sets of rules are also organized as a hierarchical sequence; each set operates over a particular range of scales. Holling (1992), for example, describes a typical sequence for a large wading bird of the Florida peninsula. At very coarse scales, tagging records indicate that the decisions required to determine an area for an individual wading bird to locate are made over several hundred to thousands of kilometers from the bird's birthplace. Once an area is located and accepted, a home range or foraging area is established within an area covering tens of kilometers. Within that area, smaller habitats are identified and exploited among a set of ponds of various sizes; within those ponds, still smaller patches of food aggregation are selected; and within those patches, specific types and sizes of food items are identified. Each of these elements has a turnover time that correlates with its geographic size. Based on sufficient data from enough species, general equations have been developed that fix the spatial and temporal position of food, home-range, and areal choices made by animals of different sizes (Holling 1992).

The spatial range for decisions covers the same range as the ecosystem/landscape hierarchy. That is, there is a tight spatial coupling between these two hierarchies, which is precisely what one would expect if spatial and temporal discontinuities are the primary source of the discontinuities that we have detected in animal body-mass distributions. The two hierarchies do not completely overlap in time. The overall decision hierarchy operates at a speed three to four orders of magnitude faster than the overall ecosystem hierarchy. Thus, the slower dynamics of the ecosystem and the landscape largely constrain and control the variability experienced for animal decisions. And, hence, those slower ecological, evolutionary, and geological dynamics determine the discontinuous distribution of animal body sizes.

INSTABILITY, OPPORTUNITY, AND DISCONTINUITIES

Panarchies are composed of sets of processes operating at different scales. At the scales between these sets of processes, the processes controlling ecological organization are unstable. Highly variable behavior has been demonstrated for the area between domains of scale in physical systems (Nittmann, Daccord, and Stanley 1985; O'Neill et al. 1986; Grebogi, Ott, and Yorke 1987) and postulated for biological communities (Wiens 1989). These observations inspired us to examine the properties of the edges of body-mass aggregations (i.e., at discontinuities) because if body-mass aggregations correspond to scales at which organized sets of processes operate, then their edges should correspond to scales at which there is disorganization and high variability.

The edges of body-mass aggregations may be considered zones of crisis and opportunity, depending on the way a given species at these scales exploits resources and interacts with its environment, and they may be analogous to phase transitions in physical systems. Complex behaviors such as migration, nomadism, and rapid adaptation leading to speciation may evolve most efficiently and commonly at scale breaks, where there is the greatest potential reward, but accompanied by the highest potential cost. The latter proposition is more thoroughly discussed in chapter 12 in this volume.

The edges of body-mass aggregations correlate strongly with a set of poorly understood biological phenomena that mix contrasting attributes. These phenomena include invasion, extinction, high population variability, migration, and nomadism. We documented (Allen, Forys, and Holling 1999) that the body masses of endangered and invasive species in a community occur at the edges of body-mass aggregations two to four times as often as expected by chance (fig. 1.3). This association is consistent across many ecological systems. The strong correspondence between the independent attributes of population status and body-mass pattern in multiple taxa and systems confirms the existence of discontinuous body-mass distributions.

It may seem surprising that both invasive and declining species are located at the edge of body-mass aggregations. It suggests that something similar must be shared by the two extreme biological conditions represented by invasive species and declining species. An examination of the phenomenon of nomadism in birds in an Australian Mediterranean-climate ecosystem found that nomadic birds, too, cluster about scale breaks (that is, they occur at the edge of body-mass aggregations) (Allen and Saunders 2002, 2006). The clustering of these phenomena at generally predictable locations within body-mass distributions suggests that variability in resource distribution or availability is greatest at discontinuities. We also

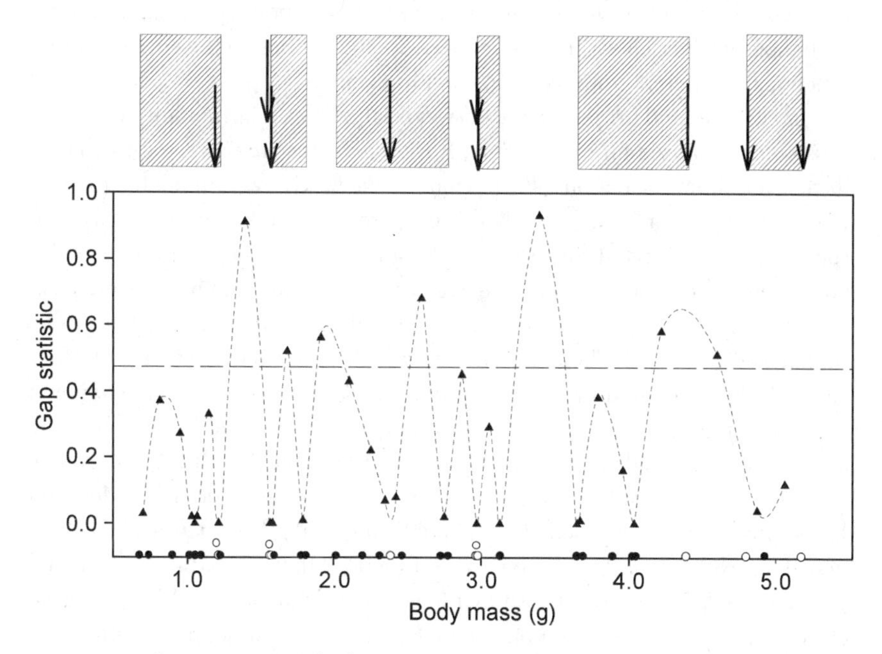

FIGURE 1.3 Gap statistic (▲), body-mass pattern (body masses as ●), and occurrence of listed species (O)for Everglades mammals. All data are presented in the lower graphic. The upper graphic displays a stylized version of the body-mass pattern and location of listed species (arrows). Aggregations (shaded) were defined as groups of three or more species bordered by significant gaps; this criterion led to us to disregard some high values of the gap statistic. Note, however, that changes in body-mass patterns due to the disregarded high values make no difference in the overall patterns detected. (Modified from Allen, Forys, and Holling 1999.)

suspect that variability in species composition and population status is higher at scale breaks (the edge of body-mass aggregations) whether or not the system has been perturbed. Extreme landscape transformations by humans, as have occurred in Florida, simply heighten the inherent variability.

It is suggestive that the most invasive of species of all, humans, had a body size on the plains of Africa that was also at the edge of a body-mass aggregation (Holling 1992). Human generalist morphology combined with gradually developed technologies allowed actions and influence at wider and wider scales—from home territories to (ultimately) the entire planet.

AGGREGATIONS IN HUMAN SYSTEMS

We have discussed ecosystems and ecological communities from a panarchical perspective, but does panarchical structure arise in other systems? A variety of research has examined the size of firms, cites, and economies. Are there clumps in the size of firms, cities, and economies? Are the entities on the edge of clumps functionally different from the entities within clumps?

For city sizes and firm sizes, the answer is largely yes (Garmestani, Allen, and Bessey 2005; Garmestani et al. 2006; further analyses are presented in chap. 8 in this volume). For nations, Barro (1997) reviews his own influential work as well as that of some others with the purpose of uncovering and measuring causal forces behind differential cross-country economic performance. He groups countries into economic aggregates that he calls "convergence clubs." Countries within a given "club" have economic growth rates that tend to be similar and converge. These patterns of growth performance across countries appear to be structured by movement toward a long-term target rate of growth for each country, wherein the long-term target is determined by slow and medium timescale variables. Slow processes of governance establish the degree of flexibility, trust, and freedom of institutional/political structures. Medium-speed processes set the general level of public physical infrastructure and education. This explanation and the nonlinear functions that support it (Durlauf and Quah 1999) appear to be quite similar to ecological panarchies. The great difficulty in moving nations from one aggregation or one development pathway to another suggests the same conservative features of clumpy patterns in ecosystems. Both nations and ecosystems seem to be sustained by conservative, slow sets of variables forming the panarchy. Both the development of nations and the management of ecosystems require that attention be focused on the slow variables, but that experiments that engage fast variables be encouraged. A critical number of the panarchy's levels need to be involved in order to satisfy minimal requirements for understanding and action.

The attraction of scaling laws is that they emerge from simple physical and statistical processes and have astonishingly wide application (Brock 1999). However, we argue that there are regular patterns of departures from or clustering along those scaling laws, and these aggregations of attributes might have more ecological, economic, and social interest and practical use than individual general laws or distributions themselves. Panarchy provides a conceptual framework for thinking about these departures.

More specifically, the existence of clumps appears to demonstrate how living systems of animals, plants, and human organizations develop self-organized interactions with physical processes over distinct ranges of scale. Just as pulses of resource acquisition over time by organisms increase efficiency of energy utilization, perhaps these "clumps" in the morphological, geometric, and behavioral variables of animals, plants, and people emerge from self-organizing properties that affect evolutionary change and development. They also have potential policy consequences. They represent attractors, created by key biological and social processes, along a more continuous, physically defined template. Thus, the measurable attributes of aggregations and discontinuities, like body-mass gaps in a distribution, are a transform of the potential for existence that is discontinuously supported across spatial and temporal scales within a panarchy.

2

SELF-ORGANIZATION AND DISCONTINUITIES IN ECOSYSTEMS

Garry D. Peterson

EMERGENCE

AN ECOSYSTEM can be characterized as an interacting set of ecological processes. The interaction of biotic and abiotic ecological processes results in the reinforcement of some processes and the attenuation of others. Ecological systems generally consist of sets of processes that produce positive feedback loops and of constraints of limits that provide negative feedback on these processes (Levin 1999).

The interaction of process and pattern at one scale produces emergent organization at larger and slower scales. Emergent processes form because of nonlinear processes acting across heterogeneous space (Nicolis and Prigogine 1977). Many small, fast processes repeatedly interact to produce a larger, slower structure that constrains the behavior of the small processes in such a way that they mutually reinforce one another. Such emergent processes are self-organized. They are not created by some outside force, but rather from the mutual reinforcement of their component processes. For example, the interaction of air and water circulation over tropical oceans can produce a self-reinforcing vortex, which strengthens itself as it draws in increasing volumes of warm air. The continued growth of this vortex produces a hurricane (Barry and Chorley 1992). Emergent processes appear to be common in ecology (Perry 1994). Indeed, one might argue that ecology is in some ways the study of how ecosystems emerge from their biotic and abiotic components.

Emergence creates new patterns. These patterns are often more robust than nonemergent patterns in that they can persist despite variation; however, when they do collapse, they do so abruptly. In this chapter, I briefly outline how self-organization in ecosystems can produce discontinuous structure and how this structure relates to the possible discontinuous organization of other variables in the system. I briefly outline the emergence of temporal, spatial, and spatiotemporal discontinuities in ecology, then provide an example of why such thinking matters.

DISCONTINUITIES IN TIME

The interactions of ecological processes over time can mutually reinforce one another to produce a specific ecosystem. These self-organized systems are frequently robust to a variety of stressors, but when change exceeds their robustness, they can suddenly reorganize, producing a temporal discontinuity in their dynamics. Such flips, or regime shifts (fig. 2.1), occur when a new positive-feedback process starts to dominate a system. A physical metaphor for this dynamic is a person sitting in a canoe. The person and canoe remain upright despite small waves and the person's wriggling, but if the person tilts the canoe over too far, positive feedback from an ever-growing tilt capsizes the canoe, reorganizing the system and throwing the person into the water (Ludwig, Walker, and Holling 1997). Shallow lakes provide an iconic example of temporal discontinuous change in ecosystems.

LOCAL EXAMPLE

Lakes can exist in either an oligotrophic or a eutrophic dynamic regime (Scheffer et al. 1993; Smith 1998). Oligotrophic lakes are characterized by low nutrient inputs, low to moderate levels of plant production, and relatively clear water. Eutrophic lakes have high nutrient inputs, high plant production, and murky water. They can also be anoxic and produce blooms of toxic algae. Oligotrophic lakes

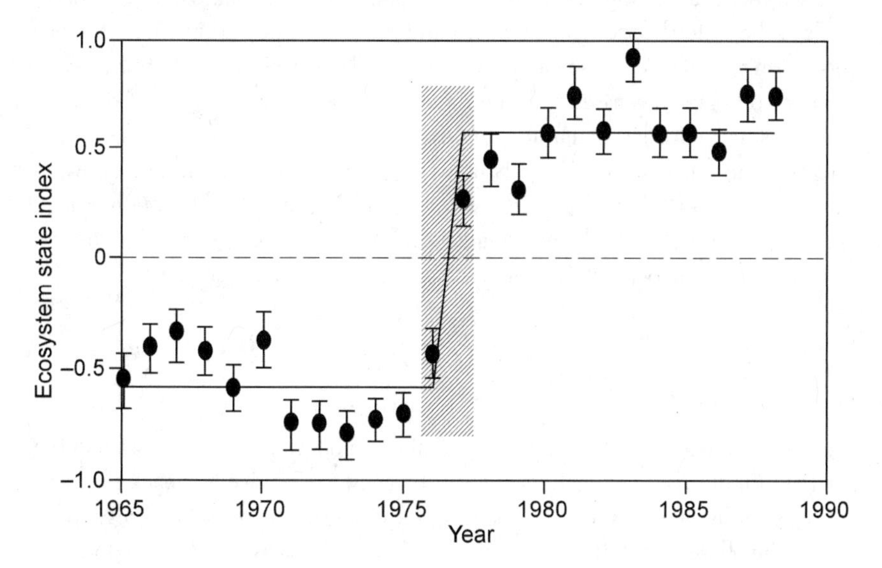

FIGURE 2.1 Regime shift in North Pacific. (Modified from Scheffer and Carpenter 2003.)

produce ecosystem services—such as water for human consumption, water for irrigation or industrial use, resources such as fish or waterfowl, and recreation—that are more valuable than those produced by eutrophic lakes (Wilson and Carpenter 1999).

Eutrophication is usually caused by the excessive input of nutrients, primarily phosphorus, to a lake (Schindler 1977). Many types of human activities produce nutrients, but agricultural fertilizers are a chief source (Carpenter et al. 1998). Phosphorus in fertilizer builds up in soil that erodes into water bodies, and it causes eutrophication (Bennett, Carpenter, and Caraco 2001).

Phosphorus recycling within a lake can maintain a eutrophic state. Recycling exhibits threshold behavior that is related to the accumulation of phosphorous in sediments, wind mixing, and the oxygen content of deep water (Carpenter et al. 1998). Experimental lake manipulations have shown that only a few years of high phosphorous levels can lead to the accumulation of enough phosphorous in the sediment to initiate phosphorous recycling (Schindler, Hesslein, and Turner 1987). In eutrophic lakes, the amount of phosphorous recycled from lake sediments may exceed annual inputs (Soranno, Carpenter, and Lathrop 1997).

The ease with which a eutrophic lake can return to an oligotrophic regime following a reduction of nutrient inputs depends on the strength of phosphorus recycling within the lake. Some factors weaken the feedback processes that maintain high levels of phosphorus, such as substantial populations of effective algal predators and macrophytes that stabilize lake sediments. Other factors strengthen these feedbacks, such as the presence of fish or waterfowl that disturb the sediments and a high concentration of phosphorous in lake sediments or the soils of the watershed. Both lake depth and the landscape within which the lake occurs shape many of these factors. Consequently, the reversibility of a transition can be comparatively assessed by considering these factors.

Transition between dynamic regimes can be found in many ecosystems that are as diverse as coral reefs (shifts between algal and coral dominance), woodlands (between forest and savanna), and arid lands (grassy to a shrub-dominated rangeland) (Scheffer et al. 2001; Walker and Meyers 2004). Such transitions are found not only at local and regional scales, but also at global scales.

GLOBAL EXAMPLES

Paleoclimatological research and simulation models have demonstrated that global climate has changed abruptly in the past as positive-feedback processes have been disrupted (Rial et al. 2004). Oceans' thermosaline circulation appears to exhibit these alternative regimes, as do feedbacks between the albedo of terrestrial ecosystems and climate, as well as between the carbon-storage terrestrial ecosystems and climate.

Thermosaline circulation is produced as salty warm water from the North Atlantic drift loses its heat in the North Atlantic, sinks, and is then slowly pushed back along the ocean bottom by more sinking water. The poleward movement of warm water, its cooling and sinking, sets up a heat- and salinity-controlled ocean circulation pattern. Consequently, this cycle is vulnerable to changes in temperature and salinity. It appears that if precipitation and runoff from surrounding land masses can supply more freshwater to the North Atlantic than is removed by evaporation, the freshened water would not be dense enough to sink, shutting down the large-scale oceanic circulation that drives substantial global transfers of heat. This shutdown may have driven large regional climate changes of as much as 8° to 16°C at the end or start of ice ages (Rial et al. 2004).

Anthropogenic global warming may be triggering large-scale global biophysical and biogeochemical feedbacks (Foley et al. 2003). An example of a biophysical feedback occurs in the boreal forest. Northern warming is allowing the northward expansion of forest into areas that are currently tundra. This expansion may decrease the albedo of northern areas, thus enhancing warming (Levis, Foley, and Pollard 2000). An example of a biogeochemical feedback occurs in the global carbon cycle. Global warming may cause the biosphere to switch from being a net sink to a net source of carbon dioxide (CO_2), further accelerating global warming. This positive feedback is due to changes in rainfall that reduce forest productivity and increase soil respiration. These changes have caused some regions to switch from being sinks of CO_2 to sources of it (Cox et al. 2000). Global examples of abrupt discontinuities demonstrate that these types of dynamics occur across all ecologically relevant scales.

DISCONTINUITIES IN SPACE

Interactions of ecological pattern and process across landscapes have long fascinated ecologists (Watt 1947). An ecological process may impose a pattern on a landscape, it may be shaped by a pattern, or it may both shape and be shaped. The intensity of the feedback between landscape pattern and the processes that shape it fundamentally influences landscape dynamics. The degree to which an ecological process is shaped by its history can be thought of as the strength of the ecological memory of that process (Peterson 2002a). In the absence of memory, there is a one-way relationship between a process and landscape pattern. The process shapes landscape pattern, but is not shaped by it. For example, if the location of a current tree-fall gap is influenced by where previous tree falls occurred, or if fire spread is influenced by where previous fires spread, then the memory of previous disturbances embedded in landscape pattern is shaping ecological dynamics in the landscape. In most ecological systems, ecological processes cannot be neatly divided between those that exhibit memory and those that do not. Even apparently one-way relationships, such as the effects of climate on vegetation, contain interactions, such as the effects of vegetation on humidity. Fire is an iconic example of how a process can both shape and be shaped by spatial pattern.

Fire is a key structuring process in many ecosystems (Bond and van Wilgen 1996). Although the duration of a fire is much shorter than the life span of the vegetation it consumes, fire produces landscape patterns that persist for long periods (Baker 1993). At small spatial scales, it homogenizes the landscape by killing above-ground growth of trees, thus producing patches of even-aged vegetation. At larger spatial scales, it produces heterogeneity by creating a mosaic of burned and unburned patches.

Fire is representative of a larger group of contagious disturbance processes—including fires, insect outbreaks, and grazing herbivores—that spread themselves across a landscape. Unlike the extent and duration of noncontagious disturbances, such as ice storms, hurricanes, and clear-cutting, the extent and duration of a contagious disturbance event are dynamically determined by the interaction of the disturbance with a landscape. The size of a contagious disturbance depends, at least partially, on the spatial configuration of the landscape being disturbed. Changes in landscape pattern will alter the nature of a contagious disturbance regime, but will not alter a noncontagious disturbance regime. For example, fragmentation of a forest by roads will impede the spread of a wildfire, but will not determine the path of a hurricane. A consequence of this interactivity is that the same driving forces will produce different contagious disturbance behaviors in landscapes with different spatial patterns.

Fire produces landscape pattern, but it is also influenced by landscape pattern (e.g., the presence of fire breaks), which in turn can also influence fire dynamics (Baker 1995). Depending on the ecological situation, fire spread can be strongly or trivially influenced by the time since an area has previously burned (Ryan 2002; Moritz et al. 2004; Schoennagel, Veblen, and Romme 2004). These differences suggest that some forests have little ecological memory, whereas others have a significant amount.

The feedback between fire spread and forest pattern produces a patch mosaic on the landscape. Relatively homogenous patches of forest are produced by the spatially contagious nature of fire spread. Memory—that is, the duration of a fire in a place—encourages the formation of mutually reinforcing patterns. When forest pattern strongly influences the spread of fires, it will also tend to be renewed by fire. When memory is weak, future fires will produce new patterns that erase former patterns (Peterson 2002a) (fig. 2.2).

Fire provides an iconic example of ecological memory, but the concept of ecological memory can also be applied to more complex ecological processes. Researchers have identified different components of ecological memory in ecosystems ranging from forests to coral reefs (Nystrom and Folke 2001; Peterson 2002a; Elmqvist et al. 2003; Lundberg and Moberg 2003). These components

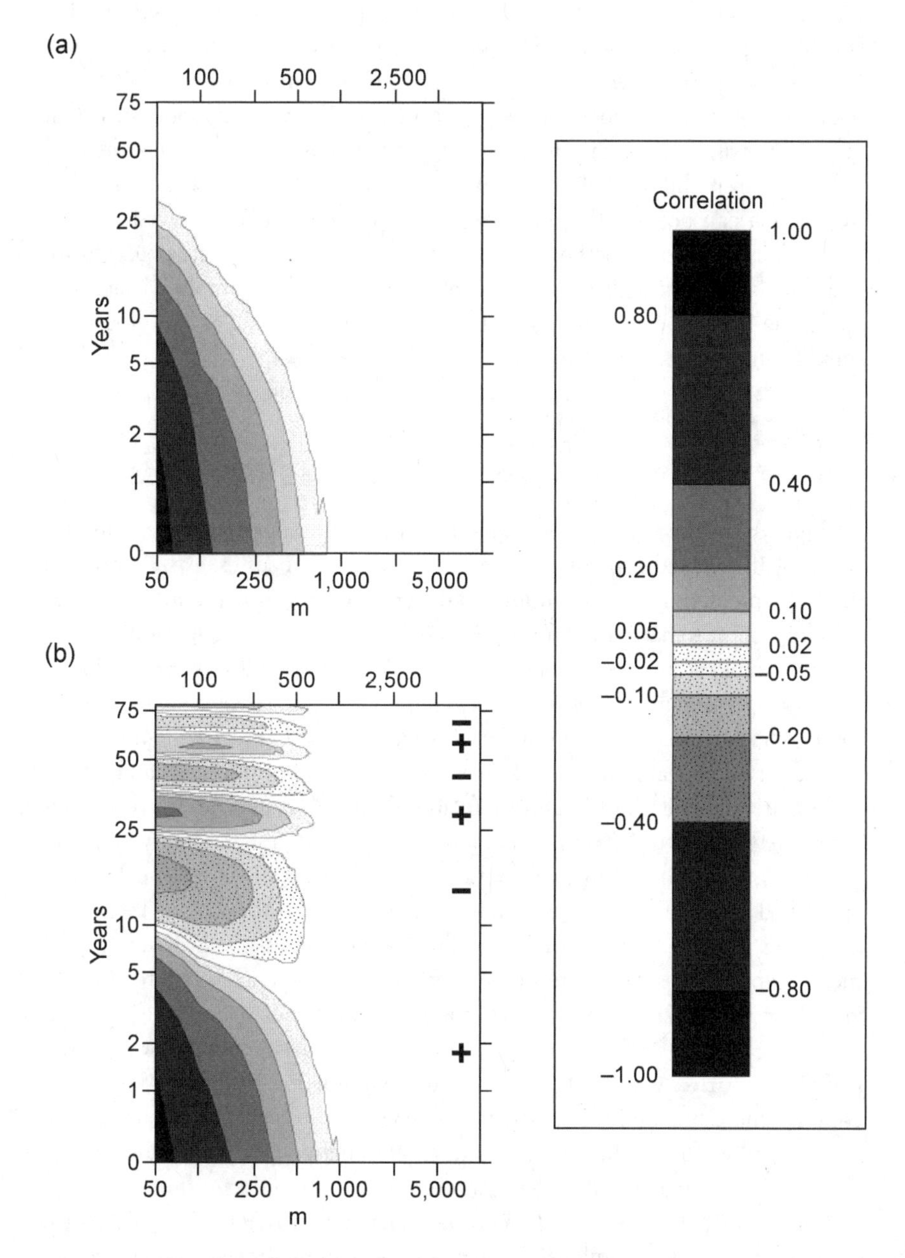

FIGURE 2.2 Creation of persistent spatial pattern with memory. Covariance at spatial and temporal lags with *(a)* minimal memory and *(b)* strong memory. Horizontal and vertical scales are log axes. Plus signs indicate positive autocorrelation, and negative signs indicate negative zones of autocorrelation. (Adapted from Peterson 2002a.)

include the physical links that connect a landscape; the mobile link species that connect together areas on a landscape; the legacies that maintain the patterns (remnants of previous ecosystems that persist despite disturbance events); and the spatial pattern that encodes previous events as well as the support areas that provide propagules, energy, and materials for the ecosystems to which they are connected by mobile and physical links. Consequently, ecological change or human action can degrade the memory present in a ecosystem, thus reducing its possibilities to produce self-organized pattern, by eliminating its potential for self-organization through the removal of legacies (by, for example, salvage logging), the reduction of functional diversity of mobile links, or the ecological simplification of support areas.

INTERACTION OF DISCONTINUITIES IN TIME AND SPACE

Multiple dynamic regimes, which produce discontinuities over time, can also coexist within a landscape that is organized by spatial processes, especially when they are interacting with environmental heterogeneity, and produce complex dynamic regimes that exhibit multiple spatial and temporal discontinuities.

An example of such an ecosystem occurs in the forests of northeastern Florida that can be dominated by either longleaf pine *(Pinus palustris)* or oaks (*Quercus* spp.) (Peterson 2002c). The ground vegetation in these forests frequently burns, and because longleaf pine and oaks have quite different responses to and effects on fire, fire mediates the competitive relationships between these two vegetation types (Heyward 1939; Rebertus, Williamson, and Moser 1989). Both young and mature longleaf pines can survive ground fires. In addition, mature longleaf pines shed needles that provide good fuel for ground fires. Young oaks are intolerant of fire. However, mature oaks shed leaves that suppress the buildup of good fuel for ground fires. Thus, fire suppression in oak stands encourages the further growth of young oaks. Without fire, oak grows up beneath longleaf pine and eventually replaces it. However, regular fires suppress oak growth, allowing longleaf pine to thrive, which allows more fuel accumulation in stands of pine, encourages more fire, and thus further suppresses hardwoods and encourages the growth of more pine (Glitzenstein, Platt, and Streng 1995). This dynamic allows for the existence of alternative dynamic regimes within this ecosystem.

Fire is a spatial process that provides connectivity across a landscape. A burning tree can ignite its combustible neighbors, spreading fire across a landscape. This ability to spread means that the spatial distribution of more or less combustible sites across a landscape strongly shapes fire spread. Under the assumption that the probability of fire ignition remains constant, sites within that area will burn more frequently if fires are able to spread across large areas than if they have difficulty spreading, which suggests that patches of longleaf pine must be larger than a minimum size if they are to burn frequently enough to prevent invasion by hardwoods (Peterson 2002b).

This example demonstrates how an ecological system may be able to exist in alternative regimes and how these regimes change when moving from a site to a landscape. In this case, a site may persist as either a young, frequently burned site (dominated by longleaf pine) or an old, infrequently burned site (dominated by oaks) (fig. 2.3). The dynamics of a single site are quite different than those of an entire landscape. When many sites in the same regime are grouped together,

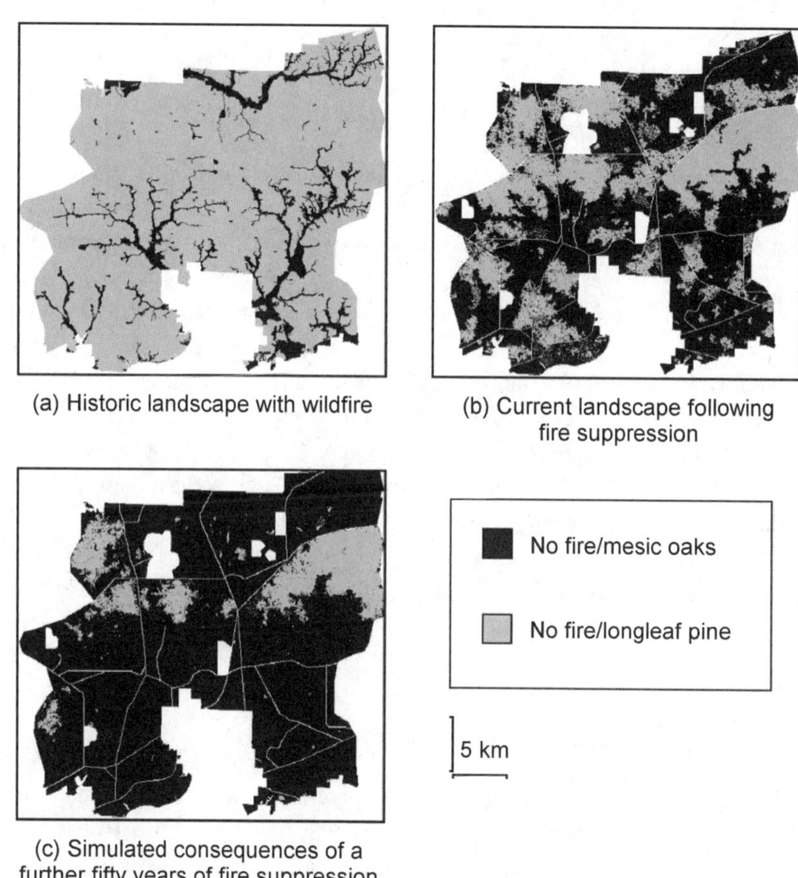

FIGURE 2.3 A schematic representation of a longleaf pine landscape (adapted from Peterson 2002b). Fire suppression has reduced the area of longleaf pine forest in north Florida. *(a)* The historical landscape was primarily a longleaf pine savanna. It was maintained by frequent wildfires. *(b)* Fire suppression and fragmentation converted a longleaf-dominated landscape to a landscape of mixed longleaf and oaks. *(c)* If historical management practices are continued, simulation models suggest that large areas of longleaf pine will be lost.

they tend to push their neighboring sites toward their type of organization by either promoting or inhibiting the ease with which fire spreads.

In the forests of north Florida, fire's ability to spread and consequently the rate at which patches of hardwood or pine either grow or shrink are determined by the distribution of hardwoods and pine across the landscape. The combination of spatial heterogeneity and positive feedbacks makes forest dynamics difficult to predict from the study of a local site because processes that define a site are determined by its neighbors' characteristics. As a result, a specific location's dynamics are strongly determined by its regional context, which means that larger-scale patterns can be used to assess the likelihood of local regime shifts (Peterson 2002b). Similar types of dynamics have been described for lakes that are connected by sports fishermen and fisherwomen (Carpenter and Brock 2004) as well as for coral reefs connected by larval coral and fish movement (Nystrom and Folke 2001; Bellwood et al. 2004).

Holling (1992) proposes that landscape pattern is generated by the interaction of a few "keystone" processes that operate at separate and distinct spatial and temporal scales. He argues that these keystone processes entrain other ecological processes and variables to their own characteristic frequencies and consequently that ecosystems' properties should exhibit discrete rather than continuous structure.

Holling and colleagues (1996) suggest that contagious disturbance processes, such as fire, are examples of such keystone processes. I argue that ecological dynamics can produce temporal, spatial, and complex spatiotemporally discrete patterns. However, these examples also suggest that such pattern requires strong feedbacks among ecological processes, indicating that Holling's (1992) Extended Keystone Hypothesis is most likely to be observed in ecosystems that have either strong competing feedbacks to produce temporal variation or strong ecological memory to produce spatial pattern.

WHY THESE IDEAS MAY MATTER

Understanding ecosystems' dynamic regimes requires understanding what drives their processes of self-organization. This knowledge matters because it allows us to assess how vulnerable current ecosystems are to unraveling, suggests methods for strengthening ecosystems, and confirms that alternative dynamic regimes are possible.

An extreme and striking example of this unraveling is provided by the Pleistocene extinctions. These extinctions occurred at different times on different continents—earlier in Africa, Southeast Asia, and Australia; later in northern Eurasian and the Americas; and later still on New Zealand and Madagascar. The observed pattern of extinctions across continents matches up with the spread of human populations, but not with climatic change (Barnosky et al. 2004). The Pleistocene extinctions predominately affected big animals. Approximately 75% of the species with body mass greater than 100 kg became extinct, approximately 40% of species weighing between 5 and 100 kg died out, whereas only a small percentage of species weighing less than 5 kg disappeared. Overall, Africa lost 18% of its mammalian megafauna, Eurasia 36%, North America 72%, South America 82%, and Australia 88% (Barnosky et al. 2004).

Pleistocene ecosystems were strikingly different from current ecosystems. For example, in the Pleistocene, North America was inhabited by five species of elephant relatives—mammoths, mastodons, and gomphotheres—as well as by saber-toothed tigers, horses, and giant ground sloths. Australia was inhabited by giant marsupials such as huge carnivorous kangaroos, marsupial lions, 7-m monitor lizards, and giant flightless birds. The Pleistocene extinctions changed the composition of communities and reduced biodiversity, but, perhaps more important, they also changed the way ecosystems functioned and reduced the number of possible ways in which ecosystems could be organized.

Large animals usually have substantial impacts on ecological functioning. Large herbivores destroy vegetation and change the landscape's physical structure. Large predators change the numbers and, more important, the behavior of their prey. For example, elephants and fire together control transitions between forest and savanna in Africa (Dublin, Sinclair, and McGlade 1990). Large predators have been shown to have substantial indirect effects on ecosystems due to changes in prey behavior. For example, as coyotes *(Canis latrans)* entered suburban southern California, bird populations increased because coyote predation reduced the impact of housecats on birds (Crooks and Soule 1999). Similarly, the reintroduction of wolves appears to be reducing the impact of climate on the supply of food for scavengers (Wilmers and Getz 2005).

Many regions of the world, including North America, that were once occupied by large herbivores and predators for millions of years have been lacking these animals for thousands of years. Due to this long and deep evolutionary history, some conservation biologists have been wondering whether the restoration of the ecosystems before the Pleistocene extinctions would make a better goal than the restoration of those that existed prior to Columbus (Zimov 2005). They have proposed the restoration of possible keystone processes by introducing African and Asian elephants in an attempt to replace the spatial disturbance processes generated by their extinct relatives (Flannery 2002). Such proposals have many practical and ethical problems, but they raise the questions, What possible ecological communities might be produced in the presence of such species, and would it be desirable, using any set of criteria, to re-create them?

This perspective on the past is important for today because the extinction and the functional extinction of many large animals are changing the basic structure of current ecosystems. Jackson and colleagues (2001) show how simplification of coastal systems has decreased their resilience, leading to a shift from productive regimes to less-desirable regimes. Large animals often provide keystone processes that are vital in maintaining ecosystems' resilience in the face of disturbance (Peterson, Allen, and Holling 1998), but the response diversity of species, or processes, that provide keystone processes in these regimes is being eroded (Elmqvist et al. 2003).

Positive feedback between ecological processes over time and across space can lead to the emergence of temporal and spatial discontinuities across scales. These discontinuities may be produced by a few key interacting processes. Changes in biotic and abiotic drivers can cause an ecological regime to reorganize into a different regime. Understanding the processes that structure these transitions and reorganizations is a key research frontier in ecology.

3

DISCONTINUITY, MULTIMODALITY, AND THE EVOLUTION OF PATTERN

Graeme S. Cumming and Tanya D. Havlicek

DISCUSSIONS OF multimodality in body-size distributions have tended to ignore the role of evolutionary processes in shaping communities. The possible range of body forms is in part constrained by genes, which consist of existing DNA and the mutations that it is capable of undergoing. Most successful genotypes are relatively conservative. Genotypes are expressed in phenotypes, which are affected directly by environmental variables such as gravitational force, the availability of energy, the demands of cells for oxygen, and the amount of pressure exerted by the atmosphere. The phenotypes of individuals must obey both the intrinsic limits set by genotypes and the extrinsic limits set by their environment. The expression of the gene-environment compromise is further complicated by interactions between organisms, which may make plausible or existing phenotypes untenable. Consequently, the body forms of animals that compose communities in vastly different environments can be expected to differ substantially—an obvious example being the differences in body size and shape between marine and terrestrial mammals. Such differences encompass not only the total number of species and the abundance of individuals within species, but also the dominant body forms and the proportional representation of different phenotypic characteristics within a community.

Hutchinson (1959) first proposed that animal body masses, as an integrative measure of niche breadth, should follow a predictable pattern. He argued that competition should result in a regular and even spacing of body sizes in different

---

We are grateful to Crispina Binohlan and the creators of Fishbase for providing us with the body-size data and necessary conversion ratios, and to Michael Oliver for additional information on Lake Malawi Cichlids. STEVE is named in honor of Steve Carpenter, who started both of us thinking about multimodal species distributions and offered valuable comments at various stages of this study. Graeme Cumming was supported by a David H. Smith Postdoctoral Fellowship from The Nature Conservancy, and Tanya Havlicek by research funds from the National Science Foundation, the Environmental Protection Agency, and the Andrew W. Mellon Foundation. We are grateful to Steve Carpenter, Chris Harvey, Buzz Holling, John Marzluff, and an anonymous referee for their comments on earlier drafts of this paper.

size categories. Minimum and constant log-size ratios were confirmed in many subsequent studies (reviewed in Simberloff and Boecklen 1981). Although ecological theory subsequently went through a period of intense revision (reviewed in Wu and Loucks 1995), changes in ideas about the mechanisms governing species coexistence were not translated into changed expectations for body-size distributions in animal communities. More recently, Scheffer and van Nes (2006) have demonstrated the plausibility of a closely related idea: that organisms may persist simply by being different or by being the same. These authors' mathematical models suggest that niche convergence may also result in multimodal body-size distributions. However, this recent paper brushes off evolution as a structuring force and ignores previous research on the topic (e.g., Cumming and Havlicek 2002).

Based largely on studies of rodents, Brown, Marquet, and Taper (1993) predicted unimodal skewed body-size distributions for mammals, with an optimum body size at approximately 100 g (but see the critique by Chown and Gaston 1997). It was widely accepted until recently that the distribution of body sizes within a particular community should be unimodal. This view is based largely on Hutchinsonian theory and does not incorporate explicit predictions about the role of genes in determining community composition. By contrast, research in evolutionary biology has focused primarily on mechanisms that determine the body sizes of individuals or lineages (e.g., Marzluff and Dial 1991; Maurer, Brown, and Rusler 1992; Alroy 1998), instead of on determinants of the body-size distributions of coexisting species.

In a seminal paper that led to the development of this volume, Holling (1992) challenged the accepted paradigm by arguing that ecological processes operating at distinct, discontinuous scales ("textural discontinuities") can create patterns of discontinuity in the community-level distribution of phenotypic characters (expressed as a clumpy distribution of species along a body-mass axis). He supported his argument with examples of discontinuous body-size distributions in mammals and birds from several different ecosystems. Holling was not alone at this time in thinking about body-size distributions. For example, Goss-Custard and colleagues (1991) described a set of correlations between the body sizes of wading birds and their prey in intertidal ecosystems. As far as we are aware, however, Holling's 1992 paper was the first to focus primarily on deviations from the theoretical expectation of unimodality. Despite some initial debate over appropriate measures of the number of modes in a community (Manly 1996), discontinuity and multimodality in body-size distributions have now been documented in a variety of studies of animals from both terrestrial and aquatic ecosystems (e.g., Restrepo, Renjifo, and Marples 1997; Allen, Forys, and Holling 1999; Marquet and Cofré 1999; Raffaelli et al. 2000; Havlicek and Carpenter 2001; Lambert 2006).

Holling (1992) proposed an ecological explanation for the observed patterns in animal community body-mass distributions. In this chapter, we consider an alternative explanation—namely, that multimodality in body masses may arise as a consequence of the way in which evolution works. It is important to note that the ideas we present are not intended to compete with existing ecological theories, but rather to make them more complete by introducing consideration of evolutionary and historical events into the debate over the causes of discontinuities in body-size distributions.

THE STOCHASTIC EVOLUTION MODEL

The Stochastic Evolution Model (STEVE) was written to test the idea that the processes of gradual descent and extinction can produce multimodality in within-lineage body-size distributions. Although we use body size as an illustrative example, the same reasoning is applicable to virtually any continuous, species-specific character that is inherited rather than acquired. The argument follows logically from four premises. (1) The range of possible animal body sizes falls along a finite continuum; animals are unable to exist beyond certain extremes of size, but can be any size in between these extremes. (2) Most speciation events involve gradual or relatively small-scale changes, as numerous writers (e.g., Darwin 1859; Dawkins 1996) have argued; mice do not give birth to elephants, and vice versa. (3) Extinction and speciation events occur at random among species of a given lineage, but over very long time periods can be approximated using a constant likelihood. (4) No two species can have exactly the same dimensions to their respective niches (besides the issue of whether this situation can ever arise in nature, we make the classical assumption that one of the two would soon go extinct).

To incorporate these assumptions in a mathematical model, we consider the range of possible body sizes (or any other trait that differs between species) as a series of discrete compartments along a size axis. The compartments or cells may be as large or as small as the reader wishes to imagine them and need not even be the same size; the main stipulation is that each cell is occupied by only one species. At the beginning of a model run, the state of each cell (filled or empty, filled cells representing species with that trait) is set at random with a probability of 0.5. Each cell is considered in sequence in each iteration of the model (fig. 3.1). Empty cells are ignored; filled cells, which are equivalent to existing species, have the three options of stasis or extinction or speciation. The outcome is randomly determined, but the likelihood of different events is determined by the choice of speciation and extinction rates. Stasis, the maintenance of the status quo, occurs in the absence of extinction or speciation; extinction results in an empty cell; and speciation results in a second cell being filled. The assumption of gradual descent means that a vacant cell in which a new species arises will be near to the cell occupied by the parent group. Whether the newly occupied cell is immediately adjacent to the parental cell or some way removed is irrelevant for the purposes of the model, provided that there is a finite distance between them and that this range of

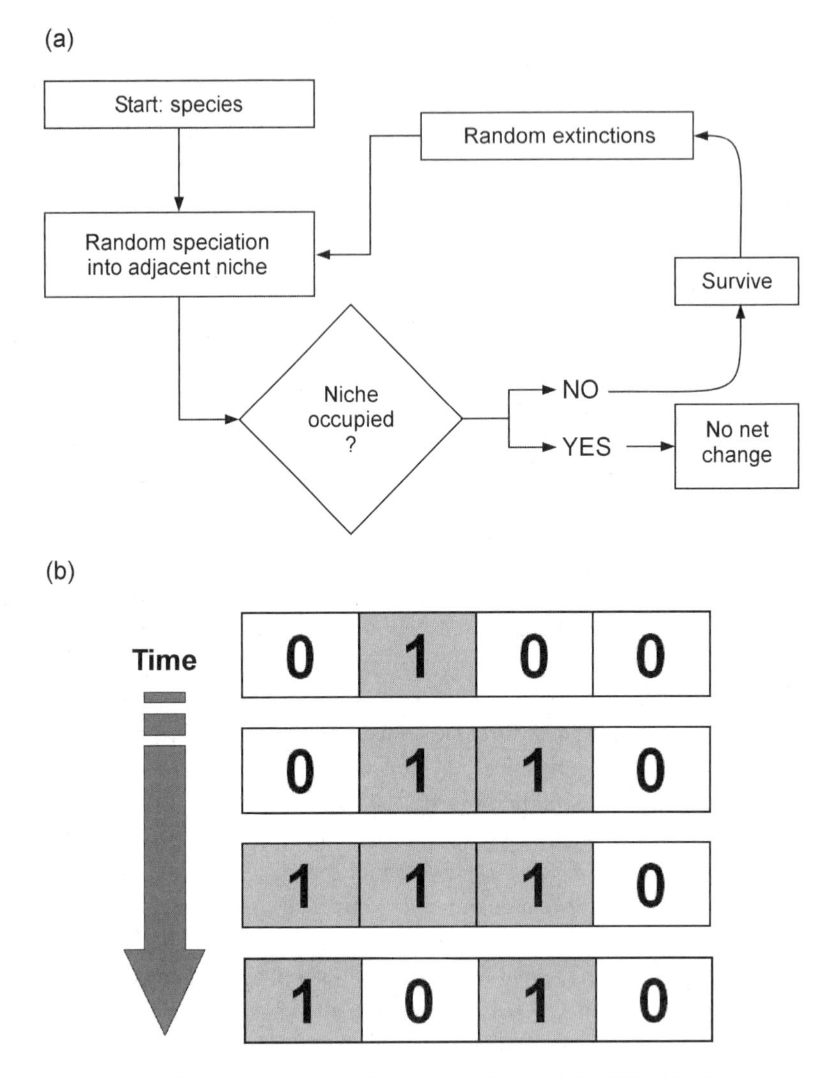

FIGURE 3.1 *(a)* Flow chart showing the steps undertaken by STEVE. *(b)* Niches are represented as cells in feature space, which may be occupied or unoccupied in any iteration. Further details are provided in the text.

variation is less than the total possible range. To keep the model simple, we have assumed that speciation can occur only into a directly adjacent cell.

When the opportunity for divergence arises, successful speciation will occur only if one of the adjacent cells is vacant. The cell to the right of the parent cell is considered first; if this cell to the right is occupied, the cell to the left is considered (the right-left order, based on Cope's rule, is irrelevant to the model's function). If both cells are occupied, no speciation event takes place. The process of evolution can in this way be described as a time series of frames, where each frame consists of a linear array of cells. The number of cells was set at one hundred throughout the analysis. Trials in which the array was "wrapped" (meaning that exchange could occur between the two end cells, creating an infinite space) to eliminate the possibility of standing waves and other edge effects yielded results that were indistinguishable from those in which the array was "unwrapped."

Readers who are familiar with explicitly spatial population models will recognize the form of this model as a stochastic cellular automaton (Tilman and Kareiva 1997). The properties of such models have been thoroughly explored and have yielded some interesting conclusions about the possible distributions of organisms in homogeneous environments. Chief among these conclusions are the predictions that no species living in a spatially limiting environment can occupy all sites at equilibrium, and a species living in a homogeneous but spatial habitat will take on a spatially clumped distribution, raising the possibility that the null hypothesis for the distributions of animals in space should be one of clumping (Tilman and Kareiva 1997). These predictions imply, in this context, that no single lineage can occupy the full range of possible body sizes and, more interesting, that across a range of related species the null hypothesis for the distribution of body size (or of any other continuous taxonomic character) should be one of multimodality. This suggestion is of particular interest in the context of the debate over the mechanisms behind body-mass discontinuity and multimodality because it indicates a mechanism for the generation of multimodality that is independent of environmental heterogeneity.

Each STEVE simulation was run for fifty thousand iterations. We used only the last forty thousand simulations in the analysis so that the (random) starting conditions did not bias the results. For example, even at relatively high extinction levels, it would take many iterations for the species present in year 1 to be removed from the simulation. The large number of simulations required by the analysis (and the large amount of data generated by a single simulation) created computational problems for the identification of multiple modes in the data, making it unfeasible to use a kernel density estimator (as in Havlicek and Carpenter 2001). Instead, we quantified the model's output by examining how the occupancy of individual body-size cells changes over time, and we tested for multimodality rather than for discontinuity per se.

Human perceptions of multiple modes are considered to be at the scale of a single time frame in the model. Ecological mechanisms proposed for the development of multimodality are based largely on extant species, which means that they are derived from a single snapshot in time. If we could take multiple snapshots through time, the persistence of body-mass multimodalities would be easier to test. If we reason in reverse, then given multiple snapshots by STEVE, the probability that multimodality would be detected in any extant ecosystem is equivalent to the proportion of time frames in which multiple modes occur. This perception is not dependent on constancy in the position of modes through time. Differences in the cumulative occupancy of niches over time offer a way to evaluate the potential for multimodality; multiple modes are more likely to be observed when some niches are frequently occupied and are interspersed with others that are frequently vacant. By contrast, an evenly distributed total occupancy of niches over a long time period would indicate greater plasticity in the distribution of modes and a lower likelihood of detecting multiple modes in any one time frame. With the single exception of a static, unimodal distribution, the likelihood of detecting consistent multiple modes in a body-size distribution over time correlates with the difference in total occupancy over time between the niches with the highest occupancies and the niches with the lowest occupancies. This difference is a surrogate measure for multimodality, which we term *separation distance,* and we use it throughout this chapter as a measure of the likelihood of the occurrence of multiple modes. Although use of this difference is not as rigorous as the use of a kernel density estimator, it is a logically consistent way of assessing the likelihood of multimodality for a data set of this nature. Each simulation run using STEVE was repeated twenty times for each combination of extinction and speciation rates, and the mean separation distance and its standard deviation were calculated.

RESULTS FROM STEVE

As predicted by cellular automaton theory, STEVE results in a clumping of occupied cells in body-size space. Depending on the model parameters and the stochastic outcome for divergence and speciation, individual frames may at any one time show the full range between "all cells occupied" and "all cells empty"; a single, stable equilibrium is not reached unless all species go extinct. The likelihood of multiple modes arising under the assumptions of stochastic evolution is dependent on both the relative and the absolute rates of speciation and extinction (fig. 3.2); it is highest when both speciation rates and extinction rates are low, and the extinction-speciation ratio is approximately 0.47. The variation in possible outcomes is also greatest at this level. When extinction is greater than 0.5 times speciation, extinction of the lineage becomes inevitable. When extinction is very low relative to speciation, the likelihood of multimodality decreases. This situation is most likely to produce the classical body-size continuum postulated by Hutchinson (1959). As the absolute values of extinction and speciation increase, species move in and out of niches faster, and the composition of body-size modes becomes less stable.

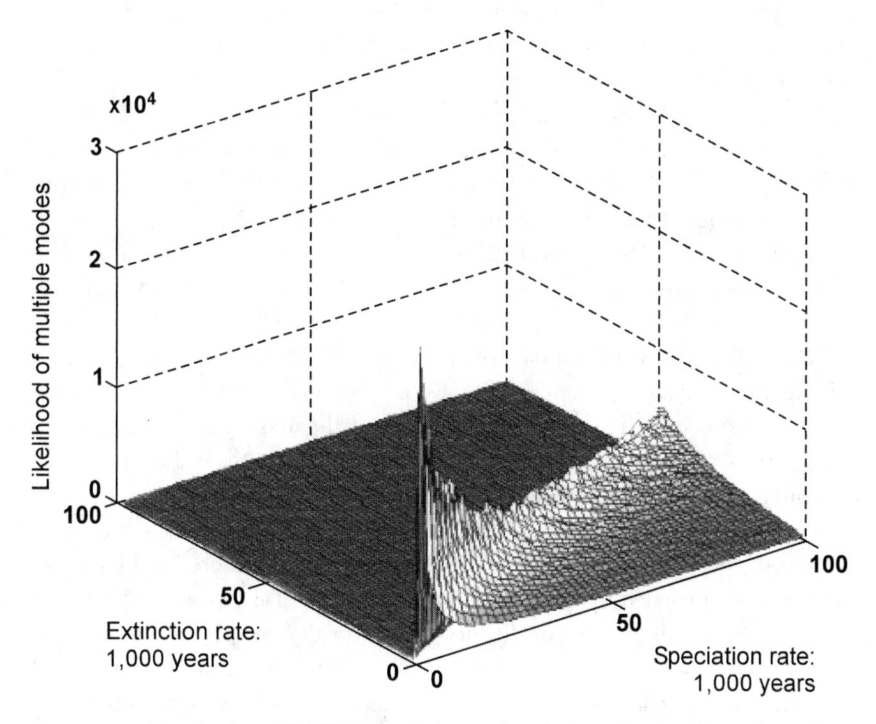

FIGURE 3.2 Results from STEVE simulations of evolution showing the relationship between extinction rate (E), speciation rate (S), and the likelihood of multimodality arising within a taxon. The ridge falls at approximately E/S = 0.47.

A REAL-WORLD TEST: FISH BODY-SIZE DISTRIBUTIONS

Because fish exhibit indeterminate growth, their upper size limit should present a strong case for the Textural Discontinuity Hypothesis (Holling 1992) in that fish generally grow as large as their environment permits (Diana 1995). If environmental limits on growth are discontinuously distributed, they should be reflected in the body sizes of fish even more clearly than in the body sizes of organisms with determinate growth.

Distinguishing between textural discontinuity and stochastic evolution is difficult; evolution and habitat change are intimately linked (Mayr 1976). Nor are ecological and evolutionary hypotheses mutually exclusive of one another; they explain the same phenomenon at different levels of explanation. STEVE predicts that faster-speciating lineages should have fewer body-size modes. More-speciose genera are likely either to be older or to have faster speciation rates and/or lower extinction rates than less-speciose genera (Greenwood 1981). Rapid speciation is more likely to occur in a complex environment with multiple niches and high spatiotemporal heterogeneity. Textural discontinuity similarly predicts that animal communities from complex habitats will have more size modes in their distributions than will those from simple habitats. This hypothesis is difficult to test directly, for several reasons. Besides the difficulties associated with measuring habitat complexity, it is not clear how speciation opportunities and constraints on speciation (e.g., competition or parasitism) balance in more complex landscapes. Although the results from STEVE are independent of assumptions about habitat heterogeneity, the mechanism of natural selection relies on both internal generation of variation and the influence of the external environment on individual survival.

Using global body-length data obtained from Fishbase (2000), we tested the prediction that larger genera should have proportionally fewer body-size modes than smaller genera. The variable of interest was the maximum known standard length for each species, regardless of its habitat. All lengths were expressed as maximum observed standard length (SL; measured in centimeters from the tip of the snout to the anterior end of the caudal fin) of males. If no male maximum observed body size was available, we used female maximum observed body size. The analysis included data for 5,823 species of fish in 138 genera. These genera represent nearly all genera globally for which more than 11 species have been described and measured. We chose a cutoff of 11 species per genus because we felt that analyses of multimodality on smaller groupings would be inconclusive. Our use of a taxonomic criterion rather than a community criterion means the species that compose each group do not necessarily occur in the same habitat or geographical region. Geographical influences may have affected the data through the location of the maximum observed size and local rates of speciation, distribution, and extinction; spatial variation would create noise in the data but is unlikely to have introduced any systematic bias, especially given the large sample size and global nature of Fishbase (2000). We did not consider the geographic location of maximum observed length or range in body size in any of our analyses.

The number of body-size modes for each genus was measured using kernel density estimation. We assumed a standard deviation equal to the larger of either 5% of the largest body size in the genus or 2 cm. This assumption allowed us to analyze distribution structure without using predetermined size classes or bins to estimate the number of modes in a distribution, thus sidestepping many of the theoretical and statistical criticisms of prior analyses to detect clumping or discontinuity. The kernel density estimator was:

$$\hat{f}(t,h_i) = \frac{1}{nh_i}\sum_{i=1}^{n}\varphi\left(\frac{t-x_i}{h_i}\right),$$

where $\hat{f}$ is the kernel density estimate, $\varphi(y)$ is the standard normal density $(1/\sqrt{2\pi})$ $\exp(-y2/2)$, n is the number of species in a genus, t are evenly spaced points along the size axis at which density is estimated, x_i is the maximum observed SL of each species, and h_i is the standard deviation associated with each x_i (Silverman 1986; Efron and Tibshirani 1993). Selecting a normal distribution to approximate the maximum size distribution of individual species in kernel density estimation gives a smoother aggregated genus distribution than would result from multimodal or skewed distributions because normal species' densities contribute more evenly to the overall summed distribution.

A critical component in the construction of kernel density functions is the choice of h_i, also called the *window width* or *smoothing parameter* associated with each density estimate. The larger the smoothing parameter, the more even and flat each species' mass density contribution is to the overall summed distribution, and the smoother will be the kernel density distribution. This is key in interpreting kernel density functions to evaluate multimodality. We tested the sensitivity of our results to the choice of smoothing parameter value by increasing h_i to 10% and 15% of the maximum observed SL of a species, moderately large and large values for species' standard deviation. Standard error is an equally valid choice of smoothing parameter, but it would increase the resultant mode number (Efron and Tibshirani 1993); we deliberately selected methods that would give us a high chance of rejecting the hypothesis of a multimodal distribution.

We calculated the number of modes, the number of species, and the size range (SLmax−SLmin) within each genus. Mode number for each genus was determined by counting the number of local maxima in the genus's size distribution. We then created a data set of pseudogenera to investigate whether phylogenetically related assemblages of species in real genera exhibited a nonrandom pattern. For each genus size between 11 and 331 members (the same range as the real data), we randomly selected the same number of maximum observed body lengths from the real data set of 5,823 species to create a pseudogenus. This process eliminated any organization imposed by evolution in the real genera. For each pseudogenus, we calculated a kernel density estimate, mode number, and size range in exactly the same manner with exactly the same assumptions as was done for the real genera. The process was repeated thirty times for each genus size to obtain mean values of the key variables. To test the sensitivity of the results from the pseudogenera to the choice of smoothing parameter, we performed the same test of increasing h_i as used for the real genera and recalculated these outputs. Whether or not the "real" h_i is max (5% coefficient of variation [CV], 2 cm), our results and conclusions can be based on the sensitivity of the distributions to the choice of h_i and the differences between patterns in the real and simulated genera.

We analyzed the relationships among genus size, number of modes, and size range using partial correlations (Sokal and Rohlf 1981). Because the relationship of interest was that between the number of species and the number of modes in each genus, we tested this correlation while holding constant the effect of size range on mode number. The rationale was that genera with larger size ranges had more potential for multimodality by virtue of possessing more size locations, so the effect of size range needed to be nullified to clarify the presence or absence of the desired relationship. We used only the 134 genera with fewer than 123 species to analyze trend differences between the real genera and the simulated genera, excluding the two most speciose genera (*Haplochromis*, with 205 species, and *Barbus*, with 331 species) as high-leverage outliers.

FISH DATA: RESULTS

For the real genera at a 5% CV, 83 of the 138 genera had more than one mode in their size distributions, and 62 of those 83 had more than two modes. The positive relationship between size range and number of species in a genus for the real genera was much weaker than in the pseudogenera ($R^2=0.06$ versus $R^2=0.78$). For the real genera, both depauperate and speciose genera can have either large or small size ranges. By contrast, in the pseudogenera, speciose genera have larger size ranges than depauperate genera (fig. 3.3). There was a strong positive relationship between size range and the number of modes for both the real genera and the pseudogenera (fig. 3.4), although the pseudogenera showed a smaller range in both quantities. The average number of modes in the pseudogenera was

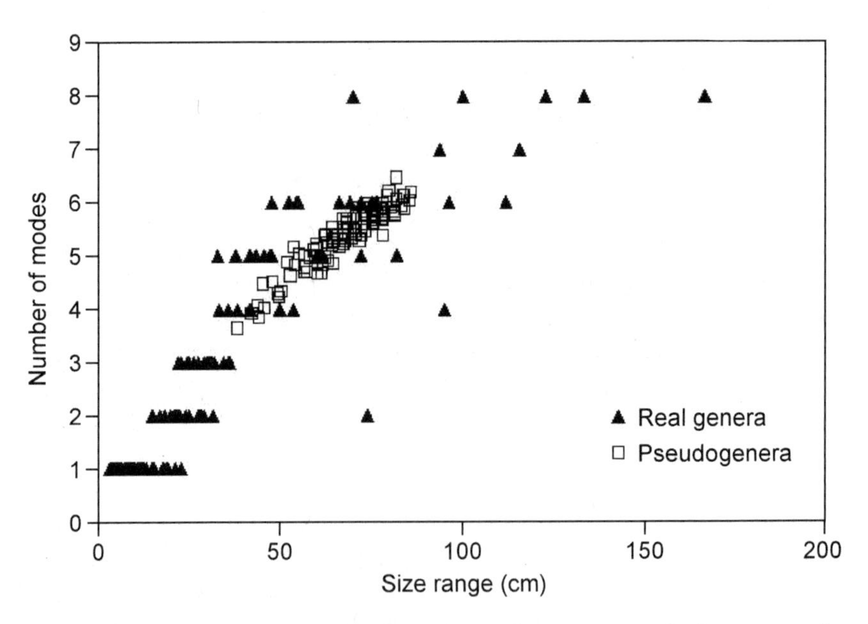

FIGURE 3.3 Plot of the number of modes within each genus against body-size range by genus for real genera (▲) and pseudogenera (□).

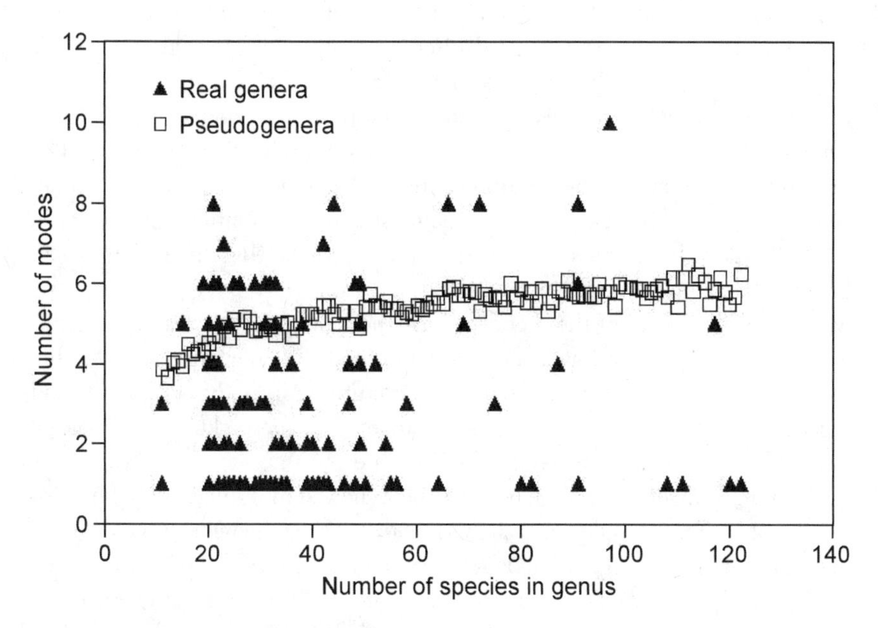

FIGURE 3.4 Plot of the number of modes within each genus against the number of species in the genus for real genera (▲) and pseudogenera (□).

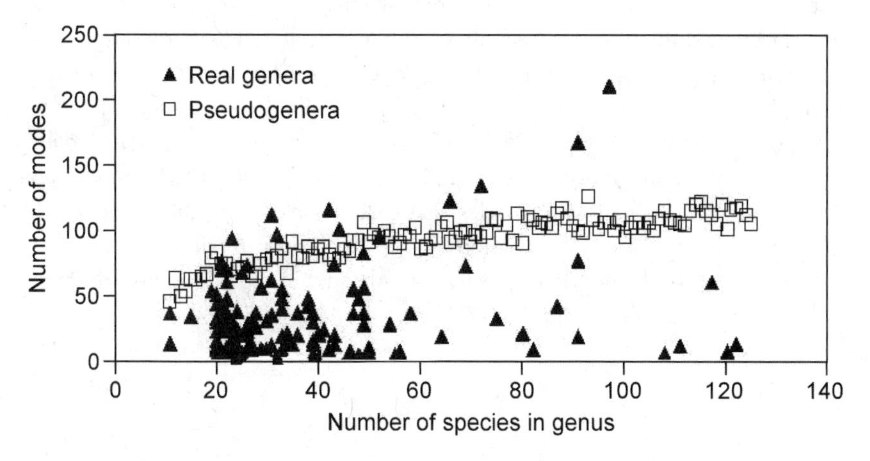

FIGURE 3.5 Plot of the size range within each genus against the number of species in the genus for real genera (▲) and pseudogenera (□).

6.2, and the mean standard deviation of mode number for each genus size was 1.3. As genus size increased, the pseudogenera increased in mode number within a tight range of values, whereas the real genera displayed a more diffuse pattern (fig. 3.5). The real genera tended to exhibit fewer modes than the simulated data.

Partial correlations between the number of modes and the number of species in each real genus (compensating for differences in size range) yielded a negative and significant relationship (partial Tau = −0.25, P = 0.03). In other words, once the relationship between size range and mode number is corrected for, bigger genera have proportionally fewer body-size modes than do smaller genera. The same partial correlations between mode number and genus size in the pseudogenera showed a significant, positive relationship between number of species in the genus and the number of body-size modes (partial Tau = 0.21, $P = 0.02$). If genera were simply assembled at random from a range of possible species, one would expect that (as for the pseudogenera) the number of species in the genus would correlate positively with the number of modes. The opposite trend in the real genera suggests that evolution plays a structuring role in the distribution and organization of body sizes by tending to smooth the more clumpy template of random aggregations as species number increases.

Both the real genera and the pseudogenera showed multimodality. The sensitivity analyses for the real genera showed that 30 of the 83 genera with more than one mode did not change in mode number at a 10% CV, and 15 of the 83 genera did not change at 15% CV. At the 10% increase, only 4 of the 83 genera became unimodal, and at the 15% level 30 of the 83 genera became unimodal. The procedure is thus relatively insensitive to the choice of smoothing parameter, a minor point given that the same methods for assessing modality were applied throughout and that the results are based on comparisons between the real genera and pseudogenera rather than on absolute values for each individual grouping.

Interestingly, the large African cichlid genus *Haplochromis*, a well-documented example of a rapid, recent speciation, showed significantly fewer modes than expected by random assemblage. There are 205 species in the genus *Haplochromis*, with a mean size range of 85.07 cm and two modes. Pseudogenera composed of 205 species had a mean of six modes (standard deviation 1.11) and a mean size range of 89.85 cm (standard deviation 17.72). Thus, the size range for a genus of 205 species was the same whether real or simulated, but the real genus was much smoother in its size distribution.

DISCUSSION

The results provide both theoretical and empirical evidence that an evolutionary perspective should be incorporated into current mechanistic theories of body-size distributions. Explanations on an ecological timescale are valid, but not complete. STEVE suggests that the existence of multiple modes in the distribution of body size and other characters can be explained as a simple consequence of the way that evolutionary processes operate and, in particular, by the two cornerstones of natural-selection theory, gradual descent and competition.

Regardless of whether STEVE is a realistic simulation of evolution, it shows that multimodality in body-size distributions within lineages can be generated from a few very simple mechanisms. The body-size distributions of fish provide empirical support for the hypothesis that the evolutionary process can produce structured, nonrandom body-size distributions.

Some adjustment to current null hypotheses may be required. Our results suggest that multimodal body-size distributions should be expected in many circumstances and that the discovery of a character that varies along a smooth, unimodal continuum for a range of related species would be peculiar rather than expected. STEVE predicts that unimodality can occur, but only under exceptional circumstances where speciation is rapid and where few or no extinctions have occurred. This conjecture is supported by the results for *Haplochromis*, which is known to have radiated rapidly with relatively few extinctions and shows significantly fewer body-size modes than expected.

To move from distinct lineages to entire communities involves changing a hierarchical level; persistent communities are stable associations composed of representatives from multiple lineages. We can envisage several different ways in which lineage-level patterns might affect community-level patterns. In much the same way as colliding waves can be amplified or can cancel out one another, the precise nature of different lineage-level multimodalities will affect community-level multimodalities. Although our results show that evolution may produce multimodality across lineages, this finding needs to be linked to community ecology before its full impact can be assessed. Further investigation will be needed to distinguish between the roles of the ecological factors and the roles of the evolutionary factors operating at each hierarchical level and their relationship to one another, particularly given the presence of apparently constant patterns such as the entry of invasive species at the edges of body-mass modes (Allen, Forys, and Holling 1999; Allen 2006).

STEVE does not include all possible evolution-based explanations for how body-size modes might arise. An additional mechanism is suggested by what may be termed *species stacking* in body sizes during rapid radiations; if changes in size are untenable for some reason, perhaps because the niches that can be reached through body-size changes are already occupied, new species are more likely to be successful if they arise through alterations along other niche dimensions. This process will result in a taxon where competition may be minimal despite marked similarities in body size. If the simple mechanism of stochastic evolution is treated as the null hypothesis, it is difficult to see how a clear-cut test can be developed to establish whether the Textural Discontinuity Hypothesis (Holling 1992) offers an equally reasonable explanation of within-genus clumping of body sizes. Both hypotheses are consistent with the predictions that multiple modes in body sizes may persist through time with relatively constant centers. Running a simulation model may render many possible scenarios; reality offers us a sample size of one. Because of the breadth of outcomes that the Stochastic Evolution Hypothesis can generate, it is difficult to refute.

Textural discontinuities should be easier to test for with empirical data, given an appropriate range of contrasts, because the Textural Discontinuity Hypothesis predicts a direct link between the habitat structure, on the one hand, and the size and nature of body-size modes, on the other. Palaeontological data that shows shifts in body-size mode centers through time, within a relatively constant environment, would probably be adequate grounds for rejecting habitat architecture as the prime cause of multimodality in body-size distributions. Empirical tests from the fossil record are complicated by the fact that environmental disturbances often drive speciation and extinction processes (Mayr 1976); changing environments are important to STEVE, while also initiating a restructuring of the processes that the Textural Discontinuity Hypothesis would rely on to produce body-size modes. Ultimately, the question may be best resolved through simulation models that explore the consequences of each set of assumptions independently and in relation to one another.

Both ecological and evolutionary factors offer feasible explanations at different scales for the same pattern, and the two hypotheses are by no means mutually exclusive. In fact, in many of the tests that we could think of to try to distinguish between their consequences, we found that the two theories are mutually reinforcing. Although evolution operates at a slower frequency than ecological processes, it is not necessarily primary; each can constrain the other depending on the specific circumstance. Holling's (1992) original conjecture was that resonance among processes operating at different scales should create a clumpy pattern in animal distributions. Although this may well be the case, multimodality in body-size distributions does not on its own demonstrate the existence of textural discontinuities in the environment. No theory of body size can be considered adequate unless it takes evolutionary history into account.

4

DISCONTINUITIES IN BODY-SIZE DISTRIBUTIONS

A View from the Top

Pablo A. Marquet, Sebastian Abades, Juan E. Keymer, and Horacio Zeballos

BODY-SIZE DISTRIBUTIONS have become a major focus of research in ecology because they seemingly reflect the operation of fundamental principles underlying the otherwise idiosyncratic nature of ecological systems (e.g., Hutchinson and MacArthur 1959; May 1986). They have been analyzed at different scales of space and time, from local communities (e.g., Brown and Nicoletto 1991) to continents and the biosphere (Blackburn and Gaston 1994a, 1994b), and from millions of years in the past (Jablonski 1996). At the landscape scale, it has been hypothesized that they reflect the existence of fundamental discontinuities in the temporal and spatial distribution of resources within ecosystems (i.e., the Textural Discontinuity Hypothesis, Holling 1992) that provide windows for persistence and invasiveness of species along the size spectra (Holling 1992; Lambert and Holling 1998; Allen, Forys, and Holling 1999; Allen and Saunders 2006).

It is not clear how the Textural Discontinuity Hypothesis, notwithstanding its theoretical appeal and empirical support, can be connected and reconciled with large-scale patterns and their explanations. As we outline in this chapter, there are several complications to performing such as synthesis across scales. Our aim is not to propose a solution to this problem, for that goal is still beyond our reach, but to present a framework to put the problem in perspective. To do this, we focus on some of the explanations put forward to understand discontinuities and aggregations in body-size distributions at continental scales.

It is customary to start a paper on body size by citing the work of Peters (1983), Calder (1996), and Schmidt-Nielsen (1984) while making the point that an organism's size and statistical distribution are of paramount importance to understanding ecological as well as evolutionary patterns and dynamics across scales in time, space, and levels of organization. However, as we aim to underscore in this

We acknowledge support from Grant FONDAP-FONDECYT 1501-001 to Centro de Estudios Avanzados en Ecología y Biodiversidad. This essay is contribution number 5 to the Ecoinformatics and Biocomplexity Unit.

chapter, body size is in a fundamental way a double-edged sword, a blessing and at the same time a curse, a sunny day and a storm.

It is safe to say that a biological system's size affects most processes that take place within it and in interactions among its biotic and abiotic elements, which is accomplished by regulating fluxes of energy and materials (e.g., Brown et al. 2004; West and Brown 2005; Marquet et al. 2005). However, this relationship does not go only one way, for the size of a system is also affected by the internal and external process it regulates, in a dance of mutual codetermination or circular causality (Hutchinson 1948; Maturana and Varela 1987; Marquet et al. 2004, 2005). The logic behind this statement is rooted in the emergent character of size. By *emergent*, we signify that the size of any biological system cannot be explained or predicted by observation of its component units (Salt 1979), but rather emerges in full complexity, novelty, and surprise as the result of the operation of many entangled processes, including how components interact among themselves and with the external environment. For the sake of generality, we can group these processes in three major categories: (1) external processes and states, related to the biotic and abiotic environment wherein the target biological entity unfolds (we use the word *biological entity* to designate any biological self-replicating system that possesses well-defined boundaries, from a subcellular organelle up to the biosphere); (2) internal processes and states, or those processes built into the entity and necessary in order to maintain its structure, function, and fitness; and (3) historical processes that describe the trajectory of the two-way interaction between the biological entity and its environment through time (the history of the interaction between internal and external processes and states). These three kinds of processes define the phase space wherein the system drifts (see fig. 4.1) or, more precisely, where we as observers make it understandable (i.e., so that the processes define an explanatory domain). In principle, any attribute of a biological entity (such as the number of components it possesses or the number species in the context of an ecosystem) can be understood, at least in principle, using any of these three major axes as a departure point (Marquet et al. 2004) and will occupy a position in this explanatory domain, which reflects the relative contribution of each of these axes to the pattern or attribute under investigation.

To see how these three axes interact in explaining patterns in body-size distributions at continental scales, we focus on the distribution of body size of mammalian species in the Americas. In figure 4.2, we show the shape of this distribution for 1,314 species of North American and South American terrestrial mammals based on data updated from Marquet and Cofré (1999) and from Smith and colleagues (2003). Interestingly, both distributions show coincident aggregations or modes for small and large sizes, but differ in a third intermediate mode apparent only in South American mammals. In the following paragraphs, we try to explain these aggregations using the three axes shown in figure 4.1.

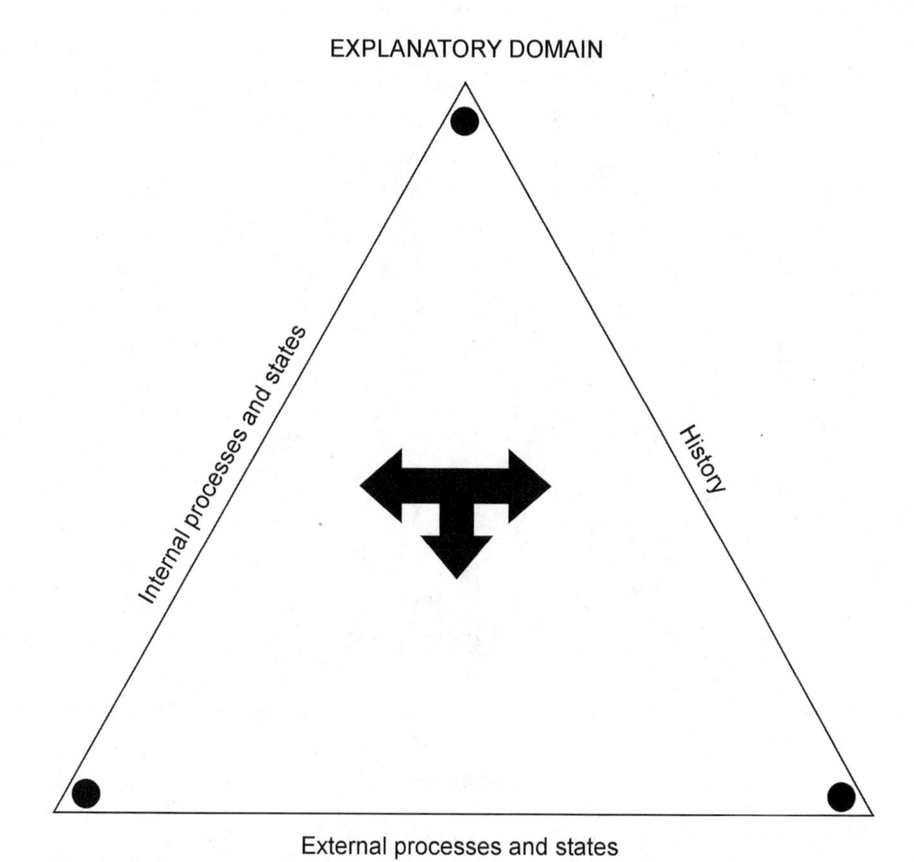

FIGURE 4.1 Explanatory domain for ecological patterns in nature. Details are provided in the text.

THE INTERNAL APPROACH

As shown in figure 4.1, the internal axes of the explanatory domain emphasizes explanations rooted in processes inherent to the organisms, which are essential to maintain their structure, function, and fitness. In this context, the explanation proposed by Brown, Marquet, and Taper (1993) provides a paradigmatic example.

Brown, Marquet, and Taper (1993) propose a model to account for the ubiquitous mode in body-size distributions of species within high-order taxa, such as mammals, birds, and reptiles. They reason that this mode reflects the evolutionary advantage of organisms of certain size given by the conflicting demands that size imposes on the rate of resource acquisition and allocation, both of which vary with size. This model has been criticized in terms of internal dimensional consistency and in terms of biological realism (e.g., Kozlowski 1996; Perrin 1998).

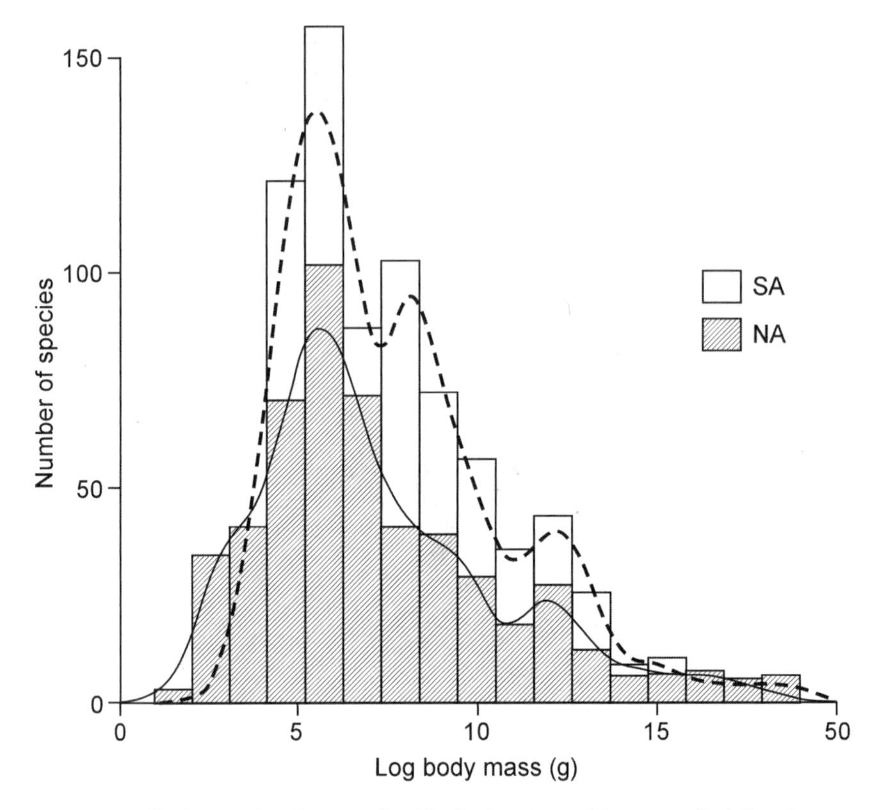

FIGURE 4.2 Body-size distributions for North American *(a)* mammals (after Brown, Marquet, and Taper 1993); *(b)* birds (after Maurer 1998a, 1998b); and *(c)* mollusks (after Roy, Jablonski, and Martien 2000). The functions correspond to the predicted distribution according to the model proposed by Brown, Marquet, and Taper (1993).

To solve these issues, we use the definitions given by Brown, Taper, and Marquet (1996) and follow the suggestions made by Perrin (1998). We sidestep the issue of biological realism and do not attempt to fit the model to real data, but rely on its qualitative behavior. The model is:

$$R + I_0 \xrightarrow{K_0} I_1 \xrightarrow{K_1} I_0 + w \qquad (1)$$

$$\frac{dw}{dt} = \frac{c_0 M^{b_0} c_1 M^{b_1}}{c_0 M^{b_0} + c_1 M^{b_1}}, \qquad (2)$$

where I_0 and I_1 are dimensionless quantities that refer to the proportion of time an individual spends in states 0 and 1 respectively. K_0 and K_1 refer to the rates of energy (resource) acquisition and transformation in units of Js^{-1}/J or s^{-1}. Thus, they are pure rates of transformation of energy either available in the environment (K_0) or already acquired and available within the organism (K_1), and are assumed to be allometric functions of body size (M) expressed as $c_0 M^{b0}$ and $c_1 M^{bl}$, respectively. Similarly equation 2 is also a rate and has the same units (Js^{-1}/J) as K_1 and K_0. The model has been applied to understand body-size distributions observed in mammals, birds, and mollusks (fig. 4.3) (Maurer 1998a, 1998b; Roy, Jablonski, and Martien 2000). Although the model's parameterization is complex and problematic (Chown and Gaston 1997; Perrin 1998; Bokma 2001), its fundamental insight is that it provides an explanation for the existence of a mode in mammalian body-size distributions based on first principles of energy acquisition and allocation in relation to reproduction (Brown, Marquet, and Taper 1993; Maurer 2003).

THE EXTERNAL APPROACH

The model makes several assumptions regarding the life history of organisms and their environment (Brown, Marquet, and Taper 1993; Kozlowski 1996; Perrin 1998). In particular, it assumes that energy or resource availability in the environment is not limited. However, ecological realism and basic thermodynamic principles inform us otherwise (e.g., Lindeman 1942). If resources are indeed limiting, then they become explicit in the equations. We can model resources as a dimensionless quantity (i.e., as a proportion of the total amount required by organisms); thus, R should vary between 0 and 1 (either no resources or plenty of resources, respectively). Under these considerations, we find that equation 2 becomes

$$\frac{dw}{dt} = \frac{Rc_0M^{b_0}c_1M^{b_1}}{Rc_0M^{b_0} + c_1M^{b_1}} \qquad (3)$$

In equation 3, it is apparent that acquisition of resources becomes a limiting process if R is less than 1; otherwise, equation 2 applies. This equation gives rise to a family of distributions, depending on the availability of resources in different environments and is shown graphically in figure 4.4. Interestingly, as resources become more limiting, the mode in body size moves toward larger sizes, thus suggesting that aggregations might emerge as a result of resource limitation (we call this the Resource Limitation Hypothesis). The reason for this result is associated with the fact that energy requirements increase with body size at a decreasing rate, and although large animals require more energy in absolute terms, they require less per unit gram, thus becoming more efficient in energy utilization (Geoffrey West, personal communication, July 2005). Thus, under resource limitation, it pays to be more efficient. In this context, if resource limitation has driven evolution toward large body size, then aggregations in body-size distributions

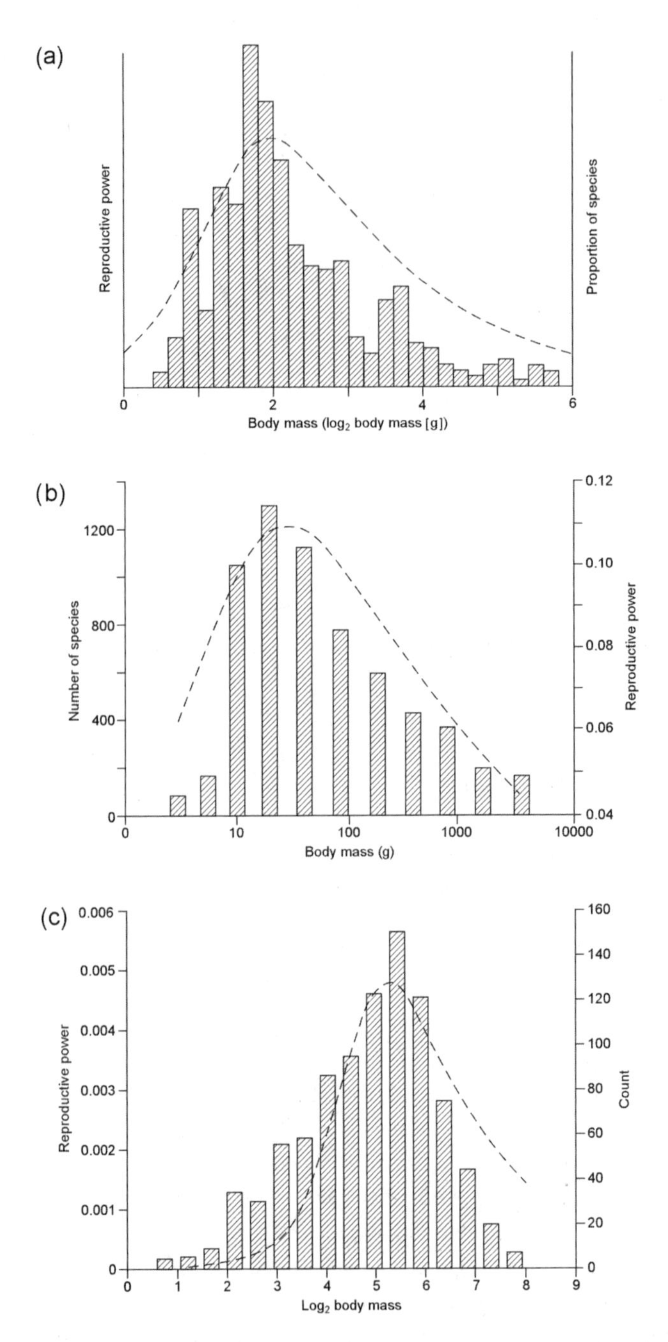

FIGURE 4.3 Body-size distributions for North American *(a)* mammals (after Brown, Marquet, and Taper 1993); *(b)* birds (after Maurer 1998a, 1998b); and *(c)* mollusks (after Roy, Jablonski, and Martien 2000). The functions correspond to the predicted distribution according to the model proposed by Brown, Marquet, and Taper (1993).

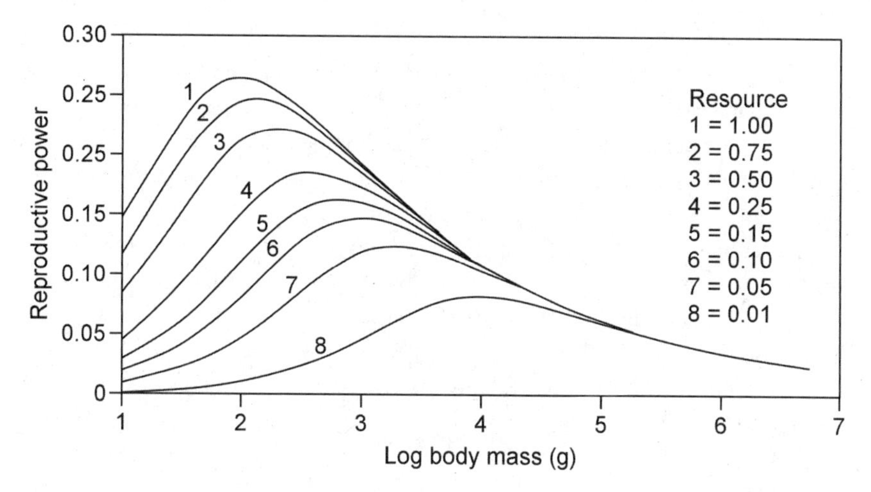

FIGURE 4.4 Predictions of the Brown, Marquet, and Taper (1993) model under different scenarios of resource limitation.

likely reflect different strategies to deal with resource shortages. Because high-quality resources tend to be rare in the environment relative to low-quality ones (Demment and Van Soest 1985), resources tend to be of lower quality as they become limiting, which for an herbivore means a higher concentration of fiber. Further, as discussed by Demment and Van Soest (1985), when body size increases, animals expand their diets to include low-quality food. In evolutionary time, this process has been accomplished by the emergence of some key innovations to deal with low-quality food, such as the emergence of gut structures that tend to delay the passage of fibrous food (e.g., rumen). Thus, the third mode observed in figure 4.2 might result as a consequence of this process. An alternative hypothesis for this aggregation in body size is based on what we call the Mode Hitchhiking Hypothesis. According to this hypothesis, it is expected that one mode can affect the emergence of other modes (i.e., through a mode-mode interaction). The most obvious case is probably represented by predator-prey interactions. In particular, the existence of empirical regularities in the relationship between prey and predator body size in ecological communities (e.g., Vézina 1985; Carbone et al. 1999; Carbone and Gittleman 2002; Cohen, Jonsson, and Carpenter 2003; Sinclair, Mduma, and Brashares 2003) may potentially emerge as aggregations in body-size distributions at large spatial scales. Thus, under this hypothesis, the existence of a large number of organisms in the 0.1 kg mode might drive the evolution of hypercarnivores of a size that will match this mode. To test for the plausibility of this hypothesis, we developed a simple model for predator- and prey-size evolution and its potential impact in body-size distributions.

THE HITCHHIKING HYPOTHESIS: A MODEL

We simulated the evolution of body mass for predator and prey species in a model that emphasized predators' dependency on the frequency distribution of available prey items. We started simulations with one hundred predator and prey species having body masses uniformly distributed in the range from 1 to 10,000 g. Body mass was always in logarithmic scale.

We let body-mass frequency distribution of prey evolve according to the following rules: extinction probability is minimal at 100 g and increases for values lower and higher than that, as suggested by Brown, Marquet, and Taper (1993). We chose to model this dependency by means of two linear regimes, where extinction probability for species weighing less than 100 g is given by

$$P_{<100\,\mathrm{g}} = 1 - 0.5 * [\text{prey mass}],$$

whereas the extinction probability for species weighing more than 100 g is

$$P_{>100\,\mathrm{g}} = -0.5 + 0.25 * [\text{prey mass}].$$

Parameter values were chosen in order to assign a probability of extinction of 1 to species tending to zero and $1 * 10^7$ grams, respectively. For every time iteration, we evaluated the extinction probability of each prey species according to the functions given here and compared the resulting value against a randomly chosen number taken from a uniform distribution $u = U(0,1)$. Whenever P was greater than or equal to u, the species in question went extinct.

We let body masses evolve by anagenesis for a random fraction of extant species, adding a random normally distributed number with mean 0 and standard deviation 0.1, as suggested by Bokma (2002). Similarly, in each time step, we allowed a random fraction of extant prey species to produce new species by cladogenesis. The rule was simply to assign a random uniform number $c = U(0,1)$ to every species and to compare it with a threshold value t (set to 0.8 in the present report). If the condition "c is greater than or equal to t" was satisfied, a new prey species was added to the extant pool. The body mass of the new species was taken from a normal distribution with mean equal to the parent species and standard deviation 0.1.

For predator species, extinction was determined by the frequency distribution of prey species. The range of prey available for a given predator was defined by the upper limit

$$ul = -1.379 + 1.819 * [\text{predator mass}],$$

whereas the lower limit was set by

$$ll = -2.618 + 2.359 * [\text{predator mass}].$$

Parameter values for these linear functions were estimated from published data on the range of prey consumed by predators (Vézina 1985; Sinclair, Mduma, and Brashares 2003). Note that *ul* and *ll* account for the fact that large species tend to prefer a broader spectrum of prey than smaller species do. Once the range of prey required by a given predator species was estimated, we scanned the frequency distribution of prey to check for the number of prey species available, A. If A equaled zero (no prey was available), the predator went extinct; otherwise, we asked for the maximum number of prey species required to sustain a predator. We modeled this latter condition as

$$M = 9.85 - 3.038 * [\text{predator mass}].$$

This function accounts for the fact that smaller predator species will require a more diverse diet than will larger ones. If M was greater than A, the predator went extinct. Anagenetic and cladogenetic changes were allowed for predators following the same rules stated for prey species.

To avoid the unbounded exponential accumulation of species, we set a limit to the maximum number of species in the system. When this limit was surpassed, we randomly removed species until the number of species fell below this level. However, to avoid distortions of the patterns generated by the evolutionary dynamics, we eliminated prey and predator species, preserving the existing ratio between them before the limit was exceeded. Similarly, we did not allow the maximum number of predators to surpass half the number of prey.

We let the system evolve for five hundred time iterations and recorded changes in the frequency distribution for both prey and predator species separately. To improve graphical display, we smoothed the frequency distributions by fitting them with a Gaussian kernel. As seen in figure 4.5, this mechanism can give rise to discontinuous, aggregated body-size distributions. Interestingly, the number of predators is the highest in the third mode (located at approximately 4kg; see fig. 4.6). However, further research is needed to assess this hypothesis and to develop more realistic models that better include the complexity associated with predator-prey interactions.

THE HISTORICAL APPROACH

So far we have been able to provide explanations based on internal and external approaches. The internal approach has emphasized the inherent constraints on organisms due to the costs of acquisition and allocation of energy for reproduction. The external approach underscores the external environment's effects in

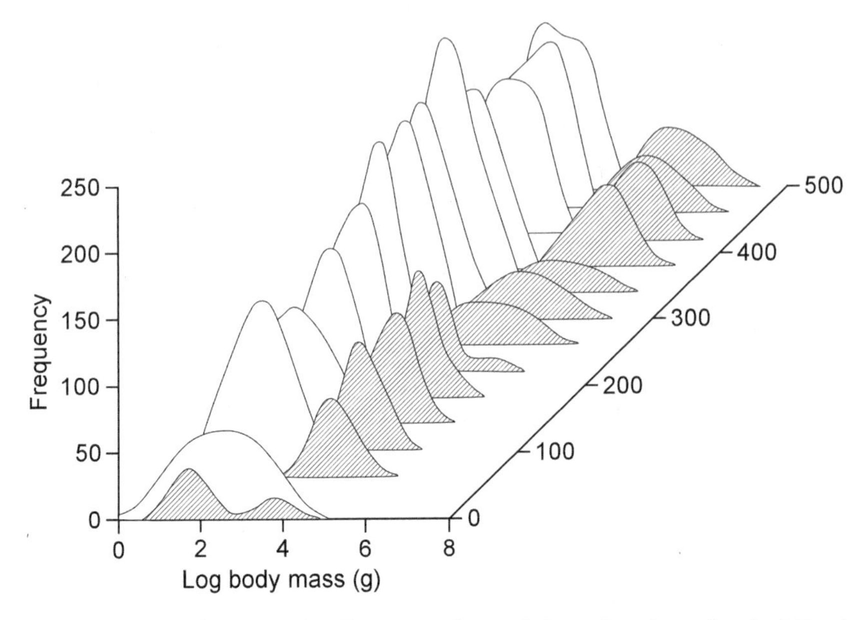

FIGURE 4.5 Simulation results illustrating the evolution of predator (hatched lines) and prey (white) body-size distributions under the Hitchhiking Hypothesis.

terms of controlling acquisition processes through resource availability. In this section, we introduce the historical approach and show that historical processes, too, may leave a mark on the body-size distribution of mammalian species in the Americas.

For South American mammals, one of the most important events that drastically changed the composition of the biota and the course of species evolution and subsequent interactions at local scales was the Great American Biotic Interchange. This event, which occurred around 2.5 million years ago, after the formation of the Isthmus of Panama, allowed the invasion of the South American continent by seventeen families of land mammals, most of which diversified and became well represented in local communities across the continent (e.g., Simpson 1980; Webb 1985, 1991; Vermeij 1991). Thus, this historical event not only altered the composition of the South American biota as a whole, but also had an enormous ecological impact in changing the pool of potential species available to assemble into local communities (Marquet and Cofré 1999). Marquet and Cofré (1999) show that for the body-size distribution of South American mammals (fig. 4.7), the first, or left-most, mode is composed primarily of species derived from North American ancestors, whereas the second mode is composed primarily of mammalian species derived from South American ancestors (mostly marsupials and Hystricognath rodents). This distinction

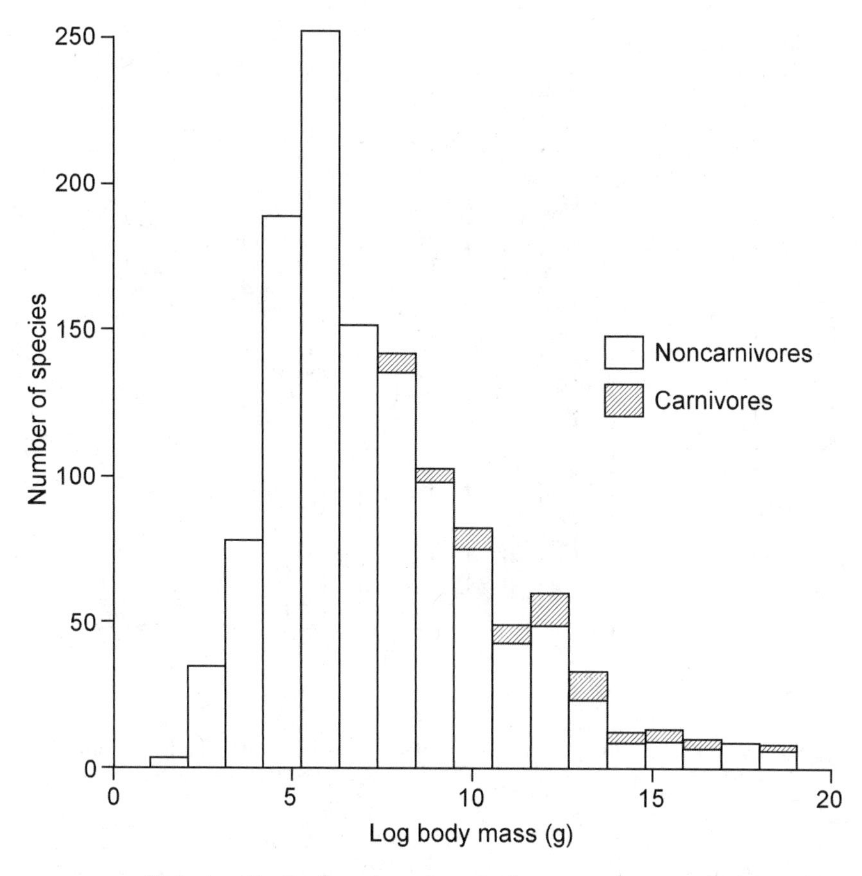

FIGURE 4.6 Body-size distribution of carnivore and noncarnivore mammal species in the Americas.

led Marquet and Cofré to propose that this bimodal distribution reflects a key event in the history of the development of the South American mammalian biota: the Great American Biotic Interchange. Marquet and Cofré hypothesize that the mode contributed by species of South American origin was also characteristic of the preinterchange continental distribution of body masses in South America and that mammals of North American origin succeeded in invading South America because of higher speciation rates (see also Lessa and Fariña 1996), which is particularly apparent in the extraordinary diversity that medium-size species (mostly rodents) of North American origin achieved in South America. This diversity might be linked to the evolutionary advantage associated with medium size (around 100g) in mammals (e.g., Brown, Marquet, and Taper 1993).

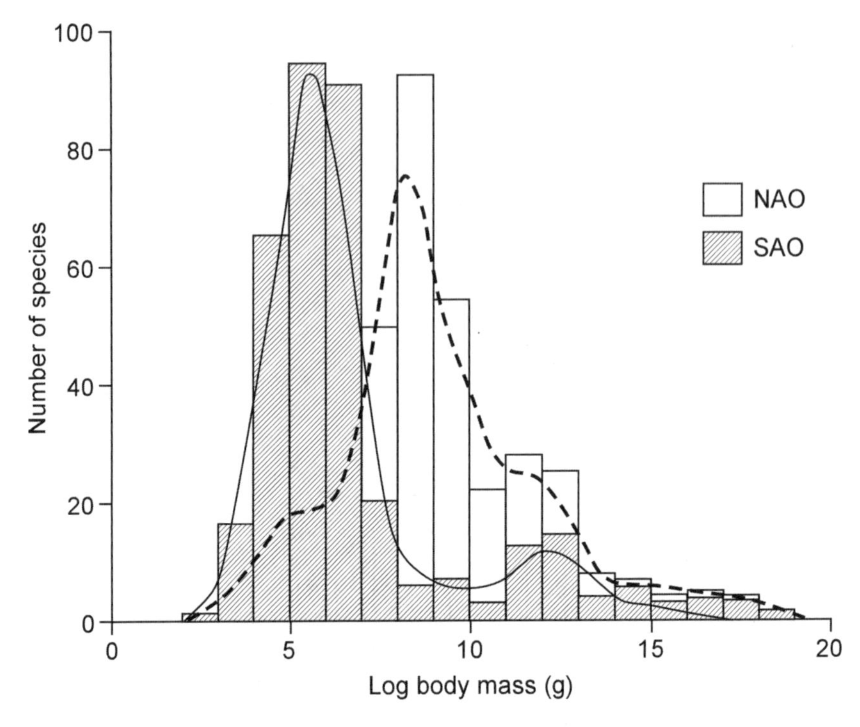

FIGURE 4.7 Body-size distribution of mammal species in South America according to their geographic origin (NAO = North American origin, SAO = South American origin).

As stated at the beginning of this chapter, our goal has been to propose a simple framework in order to put in perspective the problem of understanding discontinuous body-size distributions. As we outlined and demonstrated by analyzing the body-size distribution of mammalian species in the Americas, the main obstacle to achieving an explanation of discontinuities in body-mass distributions, at least at continental scales and for mammals, is that of multiple causality, a characteristic of ecological systems and a hallmark of their complexity (e.g., Huston 1994; Hilborn and Mangel 1997; Marquet, Keymer, and Cofré 2003; Belovski et al. 2004; Marquet et al. 2004, 2005). Ecological systems and the patterns we can discern within them reflect the history of interactions between internal and external processes and states (fig. 4.1). Consequently, they carry the signal of all these factors with different intensity and can be explained only by focusing on the explanatory domain delimited by these factors. This connection is likely more apparent when systems are analyzed at large spatial scales, such as those usually used by macroecologists (e.g., Brown 1995; Gaston and Blackburn 2000), where large-scale biogeographic patterns are thought to emerge as the result of the interactions among individual-level physiological characteristics, species' tolerances to biotic and abiotic conditions, and the large temporal- and spatial-scale process of dispersal, extinction, and speciation (e.g., Brown and Maurer 1987, 1989; Roughgarden 1989; Brown and Nicoletto 1991; Brown, Marquet, and Taper 1993; Brown 1995, 1999). It is not serendipity that most of this integration across spatial and temporal scales and across levels of organization is usually based on empirical regularities that involve the size of organisms in the form of simple scaling relationships (e.g., Marquet et al. 2005). The emergent character of body size renders it particularly suited for the task of integrating physiological, ecological, and historical factors, but at the same time makes assessing the relative contribution of each of these factors to observed patterns extremely difficult.

PART TWO

PATTERNS

5

PATTERNS OF LANDSCAPE STRUCTURE, DISCONTINUITY, MAMMAL PHYLOGENY, AND BODY SIZE

Jan P. Sendzimir

THE INFLUENCE of animal body size has been documented for a variety of biological (Damuth 1981, 1987; Peters 1983; Lawton 1989) and ecological characteristics, such as abundance (Brown 1984; Damuth 1987; Brown and Maurer 1989), population density (Damuth 1987), and home range (Harestad and Bunnell 1979; Peters 1983). Holling (1992) reversed the direction of inquiry to ask what processes influence body size. He proposed that processes such as foraging for food, shelter, and opportunities to mate may link body size with landscape structure—the size, geometry, and spatial distribution of objects in the environment—although processes related to evolution (phylogeny), trophic relations, and locomotion might also influence body-size patterns. Body size has been related to the architecture of vegetation and streambeds (Lawton 1986; Morse, Stork, and Lawton 1988; Shorrocks et al. 1991). Dubost (1979) demonstrated that artiodactyl size correlates with the structure of the undergrowth that species exploit so as to minimize resistance when moving through vegetation. Raffaelli and colleagues (2000) found that discontinuous body-size patterns in marine sediment assemblages were more strongly constrained by habitat architecture than by processes of predation or resource availability.

Discontinuities in body-mass patterns of assemblages of terrestrial animals are increasingly cited as morphological evidence of animal interactions with discontinuous landscape structure. Since Holling (1992) first identified body-mass aggregations in the bird and mammal assemblies of the Canadian boreal forest and prairie, discontinuous body-mass patterns have been identified for a variety of taxa in a number of different ecosystems. Body-mass aggregations have been shown for birds in Colombian montane forests (Restrepo, Renjifo, and Marples 1997); for birds, mammals, and herpetofauna in the South Florida ecoregion (Allen, Forys, and Holling 1999); for birds and mammals of Mediterranean-climate Australia

This chapter was greatly improved by comments from Lor Martin, Craig Allen, Craig Stow, Shealagh Pope, Garry Peterson, and an anonymous reviewer.

(Allen, Forys, and Holling 1999); for Mexican cave bats (Allen, Forys, and Holling 1999); for Pleistocene mammals in savanna forests (Lambert and Holling 1998; Lambert 2006); for lake fishes (Havlicek and Carpenter 2001); and more.

Here I explore the generality of discontinuity in animal body-mass distributions by analyzing body-mass distributions from a wide range of biomes, and I test a suite of alternate hypotheses to explain discontinuous patterns based on assumptions of links between body size and landscape structure, phylogeny, or trophic relations. I start the search to explain observations of aggregation in general and of aggregation at the simplest level: random error in sampling fluctuating populations.

ARE DISCONTINUOUS BODY-MASS PATTERNS ARTIFACTS OF SAMPLING ERROR?

If body-mass size distributions are inherently continuous, what might cause observed "lumpiness"? Failure to account for all species in a community may create gaps in a size distribution and be a source of spurious apparent discontinuity. This failure may emerge from random error in making a census, inadequate survey methods, variability in the space-time dynamics of species' populations, or some combination thereof. In the latter case, the "ephemeral" or "nonequilibrial" nature of animal communities, wherein communities' species composition appears to fluctuate unpredictably from year to year, has been the subject of lively debate (e.g., Connell and Sousa 1983; Wiens 1984; Jaksic, Feinsinger, and Jimenez 1996). Therefore, the Sampling Error Hypothesis predicts that discontinuities in body-size distributions result from random error in sampling the spatial and temporal variability of terrestrial animal populations.

DATA

I began by probing the initial work that launched much of this line of inquiry: Holling's (1992) study of birds and nonvolant mammals in the boreal prairie and forest biomes. To test the vulnerability of Holling's observations of clumpiness (discontinuity) to the Sampling Error Hypothesis in a broader geographical context, I compiled species lists ("species assemblages") of the same taxa that Holling used from seventy-five landscapes ranging in size from tens to hundreds of square kilometers from a variety of biomes on three continents (see table 5.1). I created species assemblages from official censuses at national parks as well as from published lists from wildlife research study areas. I determined a mean adult body mass for each species using values from Dunning 1993 or from Silva and Downing 1995, obtained from the geographical site closest to the landscape study site in question.

TABLE 5.1 Landscape-Study Sites Used to Provide Body-Size Distributions to Test the Sampling Error Hypothesis

MAMMAL DATA
1. BOREAL FOREST (CANADA) *site names:* Prince Albert (Doré), Nopiming Provincial Park, Taiga Biological Station, Quetico, Whiteshell
2. TEMPERATE DESERT (NORTH AMERICA) *site names:* Big Bend National Park (NP), Bryce Canyon NP, Capitol Reef NP, Canyonlands NP, Death Valley NP, Mesa Verde NP, Organ Pipe Cactus NP, Mohave Desert NP, Zion NP
3. NEOTROPICAL LOWLAND RAIN FOREST (CENTRAL AMERICA) *site names:* La Selva, Selva Lacandona, Barro Colorado Island, Sierra de Chama, Manu National Park, Alto Madidi Region, Kanuku Mountain West. Rio Cenepa, Rio Santiago
4. TEMPERATE MIXED DECIDUOUS FOREST (NORTH AMERICA) *site names:* Eglin Air Force Base, Katherine Ordway Preserve, Savannah River, Northwest Louisiana, Wind Cave, Glacier NP, Rocky Mountain NP, Yellowstone NP, Itasca
5. TEMPERATE SHORTGRASS PRAIRIE (NORTH AMERICA) *site names:* Central Plains Experimental Range (CPER), Jornada, Southwest Kansas, Pawnee
6. TEMPERATE TALL GRASS AND MIDGRASS PRAIRIES (NORTH AMERICA) *site names:* Konza, Montana Bison, South Dakota Cottonwood, Badlands NP
BIRD DATA
1. TEMPERATE MIXED DECIDUOUS FOREST (NORTH AMERICA) *site names:* Arnot Forest, Eglin Air Force Base, Finger Lakes, Hubbard, Mountain Lake Biological Station, Katherine Ordway Preserve, Turtle Mountain, Yellowstone NP
2. TEMPERATE TALL GRASS AND MIDGRASS PRAIRIES (NORTH AMERICA) *site names:* Badlands NP, CPER, Konza, Pawnee, Montana Bison, Bridger
3. BOREAL FOREST AND ASPEN PARKLAND (CANADA) *site names:* Doré Lake, Green River, Nopiming Provincial Park, Taiga Biological Station, Wood Buffalo NP
4. NEOTROPICAL LOWLAND RAIN FOREST (CENTRAL AND SOUTH AMERICA) *site names:* Alto Madidi, Ancicaya, Cocha Cashu, Guiana, La Selva, Selva Lacandona, Surinam, upper Río Urucu
5. NEOTROPICAL MONTANE WET FOREST (SOUTH AMERICA) *site names:* Cabeceras de Bilsa, Calabatea, Carpanta, Cerro Blanco, Cerro Pirre, Darién, La Planada

METHODS

ANALYTICAL MEANS TO DETECT DISCONTINUITIES I used Holling's (1992) Body-Mass Difference Index (BMDI) to examine the vulnerability of his findings to the Sampling Error Hypothesis. Discontinuities in a body-size distribution are quantified by the BMDI as the size difference between the species $n-1$ and species $n+1$ divided by the body mass of the nth species raised to a taxon-specific constant that Holling (1992) found to be 1.3 for birds and 1.1 for mammals (equation 1).

$$BMDI = [S_{n+1} - S_{n-1}]/(S_n)^{\gamma}, \qquad (1)$$

where S_i is the ith size datum in a univariate series. For the BMDI, I determined discontinuities or "gaps" in a size distribution by using Holling's criterion for a "significant" discontinuity (two consecutive BMDI values one standard error larger than the mean followed by one below that threshold value).

COMPUTER SIMULATION OF RANDOM SAMPLING ERROR I used computer simulations to test the possibility that determinations of discontinuous body-mass patterns might be the products of random events of incomplete sampling in the determination of body-size data sets. From each of the "observed" body-size distributions (table 5.1), I established a continuous, unimodal size distribution with the same range of body sizes as the observed set and created a mock data set by randomly drawing from that continuous distribution n times, where n equals the number of species in the data set. I then determined the number of discontinuities and aggregations in the mock data set using the BMDI method. I repeated this process one thousand times for each observed data set to determine mean and standard deviation statistics for number of clumps found in one thousand mock data sets.

RESULTS

Species assemblages of birds and terrestrial mammals exhibited discontinuous body-size distributions. I found three to twenty aggregations in observed bird and mammal body-size data sets, but more than 70% of mock data sets had only one aggregation—that is, were unimodal (fig. 5.1). Discontinuities and aggregations in body-mass distributions are not due to the false generation of gaps in the size distributions of species assemblages because of random sampling error. Having ruled out a relatively simple, methodological cause of discontinuity, I now address more complex ecological explanations.

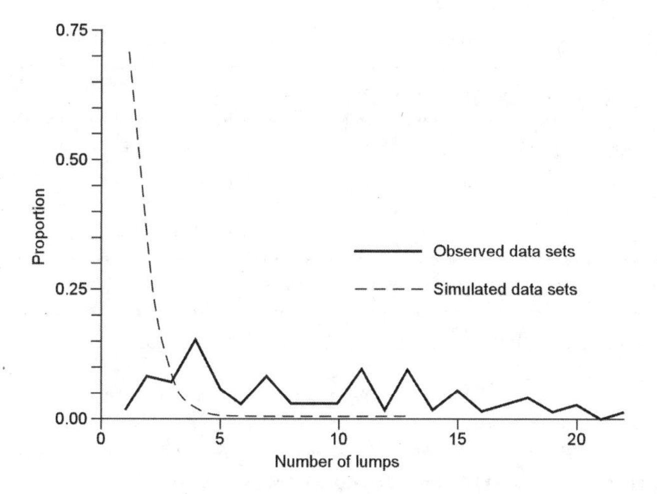

FIGURE 5.1 Proportion of lumps found using the Holling Body-Mass Difference Index on observed and simulated (mock) data sets of mean adult body masses for species of birds and nonvolant mammals. Each datum for the simulated group is a mean value of clumps found in one hundred mock data sets synthesized by random draws from the same body-size range as an observed data set.

ALTERNATIVE EXPLANATIONS FOR DISCONTINUOUS BODY-MASS PATTERNS

Demonstrations of discontinuous body-mass patterns are widespread, but an accepted mechanistic explanation for these phenomena remains elusive. In addition to the proposal that clumpiness in animal body-mass distributions reflects their interaction with discontinuous landscape structure (the Textural Discontinuity Hypothesis [TDH]), Holling (1992) posed a number of alternative hypotheses based on founder effect (Urtier Hypothesis), trophic structure (Trophic Trough Hypothesis [TTH]), and locomotory modes (Limited Morph Hypothesis) as explanations of discontinuous body-mass patterns in animal species assemblages (table 5.2). The TDH posits that assemblages of terrestrial animal species cluster into distinct body-size ranges to exploit discontinuous textures of landscape structure created by discrete and different ranges of scale at which resources are evident and available. The Urtier Hypothesis predicts that gaps are found in body-mass size distributions because limited radiations in size from each ancestor (founder) have not yet saturated the entire body-size spectrum with intermediate body sizes. This hypothesis is similar to the Core-Taxa Hypothesis (Brown 1995; Siemann and Brown 1999), and hereafter I refer to these two hypotheses as one category: Phylogeny-Based Hypotheses (PBH). The TTH predicts that gaps in body-mass size distributions result from significant size

TABLE 5.2 Hypothetical Explanations of Clumpiness in Body-Size Distributions

EXPLANATION BASED ON METHODOLOGY	
SAMPLING ERROR HYPOTHESIS	Body-size distributions are inherently continuous and unimodal, but gaps are evident because sampling error incompletely censuses the animal community.
EXPLANATION BASED ON LANDSCAPE STRUCTURE	
TEXTURAL DISCONTINUITY HYPOTHESIS	Landscapes inhabited have discontinuous textures created by discrete and different ranges of scale at which resources are evident and available.
EXPLANATIONS FROM PHYLOGENY-BASED HYPOTHESES (PBH)	
URTIER HYPOTHESIS	Organizational constraints of a limited number of ancestral forms preclude the evolution of intermediate sizes that would fill the gaps in size distributions.
CORE-TAXA HYPOTHESIS	Clumps occur because of conservation of body sizes in higher taxa over evolutionary time, which manifest as similarities in body-size distributions among biomes on the same continent and arise in two ways: taxonomic relatedness of species in different biomes and the occurrence of the same species across multiple biomes.
EXPLANATION BASED ON TROPHIC RELATIONS	
TROPHIC TROUGH HYPOTHESIS	Gaps appear in size distributions because of qualitative differences in size inherent in trophic interactions.
EXPLANATION BASED ON LOCOMOTORY MODES	
LIMITED MORPH HYPOTHESIS	Clumps result from body-size aggregations associated with a limited number of possible life forms with unique designs for moving through the environment.

Sources: After Holling 1992 and Brown 1995.

differences between predators and prey (see also chap. 4 in this volume). Given the total lack of support for the Limited Morph Hypothesis (Holling 1992; Sendzimir 1998), I confine my analysis to one broad taxon: nonvolant mammals.

In this section, I test predictions based on this suite of hypotheses in a large number of landscapes that represent a diversity of landscape structures, hydrological regimes, edaphic factors, and climates so that I might generalize conclusions about discontinuity in animal body-mass distributions beyond the limited range of factors found in the few biomes studied to date.

BODY-MASS DATA TO TEST ALTERNATIVE HYPOTHESES

I compiled species lists ("species assemblages") at two spatial scales: landscape-study sites (as previously described) and biome-level species lists. I sorted landscape-study site data sets by biome (table 5.3) (Sendzimir 1998).

TABLE 5.3 Landscape-Study Sites Used to Provide Body-Size Distributions to Test Landscape-Based (Textural Discontinuity) and Phylogeny-Based (Urtier and Core-Taxa) Hypotheses

Biome and site names
1. TUNDRA (CANADA) *site names:* Northern Yukon, Kluane
2. BOREAL FOREST (CANADA) *site names:* Prince Albert (Doré), Nopiming Provincial Park, Taiga Biological Station, Quetico, Whiteshell
3. TEMPERATE MIXED DECIDUOUS FOREST (NORTHERN NORTH AMERICA) *site names:* Wind Cave, Glacier, Rocky Mountain, Yellowstone NP, Itasca
4. TEMPERATE MIXED DECIDUOUS FOREST (SOUTHERN NORTH AMERICA) *site names:* Eglin Air Force Base, Katherine Ordway, Savannah River, Northwest Louisiana
5. TEMPERATE SHORTGRASS PRAIRIE (NORTH AMERICA) *site names:* Central Plains Experimental Range (CPER), Jornada, Southwest Kansas, Pawnee
6. TEMPERATE TALL GRASS AND MIDGRASS PRAIRIES (NORTH AMERICA) *site names:* Konza, Montana Bison, South Dakota Cottonwood, Badlands NP
7. MEDITERRANEAN (NORTH AMERICA, EUROPE) *site names:* Mckittrick, Donana
8. TEMPERATE DESERT (NORTH AMERICA) *site names:* Big Bend NP, Bryce Canyon NP, Capitol Reef NP, Canyonlands NP, Death Valley NP, Mesa Verde NP, Organ Pipe Cactus NP, Mohave Desert NP, Zion NP
9. TROPICAL DRY DECIDUOUS FOREST (CENTRAL AMERICA, SOUTH AMERICA, AFRICA) *site names:* Chamela, Bandia

TABLE 5.3 *continued*

10. TROPICAL WET FOREST
site names: Mount Nimba, Makokou, Guatopo, Columbia River

11. NEOTROPICAL LOWLAND RAIN FOREST (CENTRAL AMERICA)
site names: La Selva, Selva Lacandona, Barro Colorado Island

12. NEOTROPICAL LOWLAND RAIN FOREST (NORTHEASTERN SOUTH AMERICA)
site names: Kanuku West, Kanuku East, Surinam, Belém

13. NEOTROPICAL LOWLAND RAIN FOREST (WESTERN SOUTH AMERICA)
site names: Manu NP, Alto Río, Cerro Pato de Pájaro, Río Cenepa, Río Santiago

14. TROPICAL MONTANE RAIN FOREST
site names: Sierra de Chama, La Planada, Cuzco, Manta Real

15. TROPICAL MOIST FOREST
site names: Brazil Atlantic, Sangmelina

16. TROPICAL SCRUBLAND (SOUTH AMERICA)
site names: Caatinga, Perforación, Monte Desert, Curuyuqui

17. VELDT (SOUTHERN AFRICA)
site names: Rwindi, Transvaal 3, Transvaal 7, Transvaal 9, Golden Gate

18. TROPICAL SAVANNA WITH FOREST (AFRICA)
site names: Gabiro, Serengeti, Amboseli

19. TROPICAL WOODLAND/SAVANNA (AFRICA)
site names: Jebel Mara, Ihema, Zinave, Lokori, Rukwa, Transvaal 10

20. TROPICAL RAIN FOREST (AFRICA)
site names: Park de Tai, Lamto, Mount Kivu, LaMaboke

21. PALEOTROPICAL ASIAN RAIN FOREST
site names: Cape York (Australia), Malaysia

To examine discontinuous body-size patterns at larger scales, I analyzed body-size data sets aggregated at the biome level so as to address suspicions (Wiens 1984) that "community" data sets are too ephemeral and transient to be fundamental units of analysis. I assembled mammal species lists from five biomes that represent a range of values for landscape structure, latitude, climate, and biodiversity (table 5.4). The biome data sets include published data for boreal forest and prairie (Holling 1992) and for the Sonoran Desert (U.S. National Park Service 1998), as well as unpublished data for the Kalahari savanna

TABLE 5.4 Biome Aggregate Species Lists Used to Test the Textural Discontinuity, Urtier, Core-Taxa, and Trophic Trough Hypotheses

NO.	BIOME (CONTINENT)	REFERENCE LOCATION (FILE NAME)
1	Boreal Forest (North America)	Holling 1992
2	Boreal Prairie (North America)	Holling 1992
3	Neotropical Lowland Rain Forest (South America)	Voss and Emmons 1996
4	Sonoran Desert (North America)	U.S. National Park Service 1998
5	Kalahari Savanna (Southern Africa)	Ricardo Holdo (personal communication, Jan. 1998)

(Ricardo Holdo, personal communication, August 1998). For the Neotropical lowland rain forest, I took a subset from data sites in western Amazonia (Voss and Emmons 1996, species lists) and determined a final list following the recommendations of J. Fragoso, a researcher with considerable experience in the Amazon.

METHODS OF DISCONTINUITY DETECTION

I determined clump structure using the Gap Rarity Index (GRI) (Restrepo, Renjifo, and Marples 1997; Allen, Forys, and Holling 1999) because although both the BMDI and GRI produce very similar results, the GRI allowed me to adjust alpha levels to maintain constant statistical power levels across all analyses. Because species number *(n)* varied almost fivefold over all data sets, I maintained a constant statistical power of approximately 0.50 (Lipsey 1990) when setting alpha for detecting discontinuities (Allen, Forys, and Holling 1999).

The GRI, a methodology similar to Silverman's (1986), allows detection of discontinuities within a size distribution and comparison of their location among multiple data sets (Restrepo, Renjifo, and Marples 1997). It generates null models of species-size distribution through computer simulation using the minimal unimodal kernel estimate of data density of a species size pool. The null distribution has one mode by design in order to establish the size of gaps one would find if the actual, underlying distribution were unimodal. After ordering the observed data set by ascending size, I compared its data density to that found in the null distribution. The basis of comparison is the relative discontinuity (D_i), a statistic generated by statistical simulation of ten thousand mock distributions drawn at random from the theoretical null. Over all iterations, a distribution of gap sizes is generated at each species rank by calculating absolute discontinuities (d_i) for each pair of neighboring species (equation 2).

$$d_i = s_i + 1 - s_i \qquad (2)$$

The Relative Discontinuity (D_i) is a value found for each species that measures the proportion of absolute discontinuities calculated in all iterations of mock distributions that are smaller than the observed absolute discontinuity. This value establishes how rare a discontinuity would be in a unimodal size distribution, and this rarity is used to determine the gap's significance (alpha) level (Restrepo, Renjifo, and Marples 1997).

PAIRED TESTS OF ALTERNATIVE HYPOTHESES

TESTS OF HYPOTHESES BASED ON PHYLOGENY AND TROPHIC RELATIONS

PREDICTIONS The TTH and the PBH predict that any aggregation of body sizes in a size distribution reflects clustering of trophic groups or taxonomic groups, respectively. I tested these predictions on body-size distributions of species lists aggregated at the biome level. I tested the degree of clustering/discontinuity at two levels. I examined whether trophic or taxonomic classifications cluster at the scale of body-mass aggregations and then at scales larger than aggregations. The latter tests whether finer dissections of body-size discontinuities revealed by lump analysis obscure clustering evident at the broader scales (for example, two or three modes over the whole body-size distribution).

DATA AND METHODS First, I used the GRI to determine the discontinuities in the body-mass structure of animal data sets assembled from mammal species lists from five biomes (table 5.4). I classified each species in each assemblage based on its trophic group membership (carnivore or herbivore/omnivore), its taxonomic order, and the aggregation in which it occurred, excluding from further analysis those trophic or taxonomic classifications represented by three or fewer species. I used chi-square tests of independence to determine if membership in a taxonomic or a trophic group predicted membership in an aggregation. I set the experiment-wise significance level for the chi-square tests at 0.05 and used the Bonferroni correction (Rice 1989) to adjust individual probability values.

In the second set of tests, I examined clustering of body sizes of trophic and taxonomic classifications at scales larger than aggregations. I created body-size frequency distributions at the biome level for two different trophic groups of mammals. I assembled species-size data at the biome level by compiling body-mass data from all landscape study sites I had assembled for each biome and by removing redundant species. I then separated each biome species assemblage into two trophic groups: carnivores and herbivores/omnivores. I overlaid graphs of frequency distributions of both trophic categories for each biome.

I tested the PBH predictions by comparing two broad patterns: body-mass pattern shared across all biomes and size distributions compiled at the level of mammal orders. Through comparison of body-mass structures of ninety landscape-study sites (table 5.3) taken from twenty-one biomes, I defined where body-mass pattern is shared (i.e., the commonality in the location of discontinuities and aggregations on the size axis) across biomes by determining the zones on the log body-size axis where aggregations most frequently occur. I randomly sampled the mean adult body masses of one hundred randomly selected species of each of the seven most speciose mammal orders so that I could create an estimate of the size distribution of each taxonomic order that equally weights each order according to *n*. I then graphically compared shared body-mass and taxonomic patterns by juxtaposing the zones of highest pattern correspondence with each mammal order's size distribution.

RESULTS The five biomes chi-square tests (table 5.5) revealed no significant relationships between membership in particular-size aggregations and either of the trophic categories (carnivore/insectivores or herbivore/omnivores). The majority of taxonomic orders tested in five biomes had high probabilities that membership in aggregations is independent of membership in taxonomic orders (table 5.5). I found significant dependence between body mass and taxonomic patterns in only two cases (artiodactyls and insectivores) out of twenty-six tests (8%).

The TTH predicts that trophic interactions involving qualitative differences in size between predators and prey generate the discontinuities that separate aggregations in a size distribution. Such qualitative size ratios do not characterize the size distributions of carnivores or herbivores/omnivores in any one of six different biomes (fig. 5.2). Barring a few rare exceptions, both trophic groups coexist in any biome all along the body-size distribution.

The PBH predicts that limited and conservative radiation of body sizes creates discontinuities or aggregations in body-size distributions, or both, especially at higher taxonomic levels such as orders. The body-size distributions of these six mammalian taxonomic orders (fig. 5.3) are differentially aggregated along the size axis. Also, these aggregations coincide to some extent with the zones along the body-size axis where I found the greatest degree of alignment between clumps in different mammal species assemblages.

TABLE 5.5 Results of Chi-Square Tests of Independence Between Body Size, Trophic Level, and Taxonomic Group

	AFRICAN KALAHARI SAVANNA	NEOTROPICAL LOWLAND RAIN FOREST	CANADIAN BOREAL PRAIRIE	CANADIAN BOREAL FOREST	SUBTROPICAL SONORAN DESERT
TROPHIC CLASSES					
HERBIVORES/ OMNIVORES	0.828	0.732	0.794	0.914	0.975
CARNIVORES	0.605	0.171	0.439	0.670	0.830
PHYLOGENETIC CLASSES					
ARTIODACTYLA	0.018	0.535	0.009	0.042	0.001*
CARNIVORA	0.453	0.276	0.272	0.187	0.156
INSECTIVORA	0.007		0.000*	0.007	0.653
LAGOMORPHA			0.207		0.358
MARSUPI-CARNIVORA		0.060			
PERISSO-DACTYLA	0.280				
PRIMATES	0.855	0.473			
RODENTIA	0.011	0.042	0.133	0.291	0.036
XENARTHRA		0.831			

Note: Asterisk indicates rejection of the null hypothesis assumption of independence.

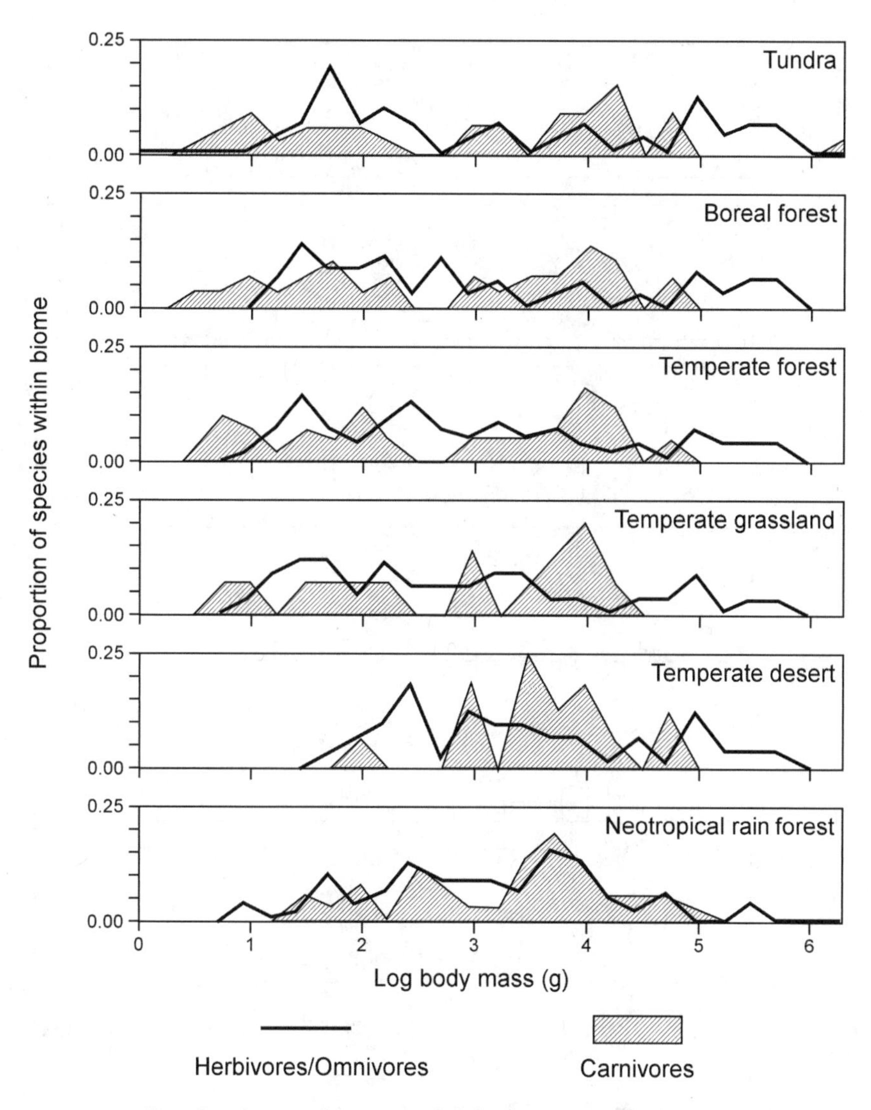

FIGURE 5.2 Size distributions for mean adult body masses of carnivores and herbivores/omnivores expressed as proportions of all species in subsamples of mammal species assemblages in six biomes.

TESTS OF THE TDH AND PBH

PREDICTIONS Both landscape-based hypotheses (the TDH) and PBH predict that similar locations of discontinuities and aggregations on the body-size axis will characterize body-mass patterns from study sites in the same biome. The basis of this pattern of similarity between sites may originate either from similarities

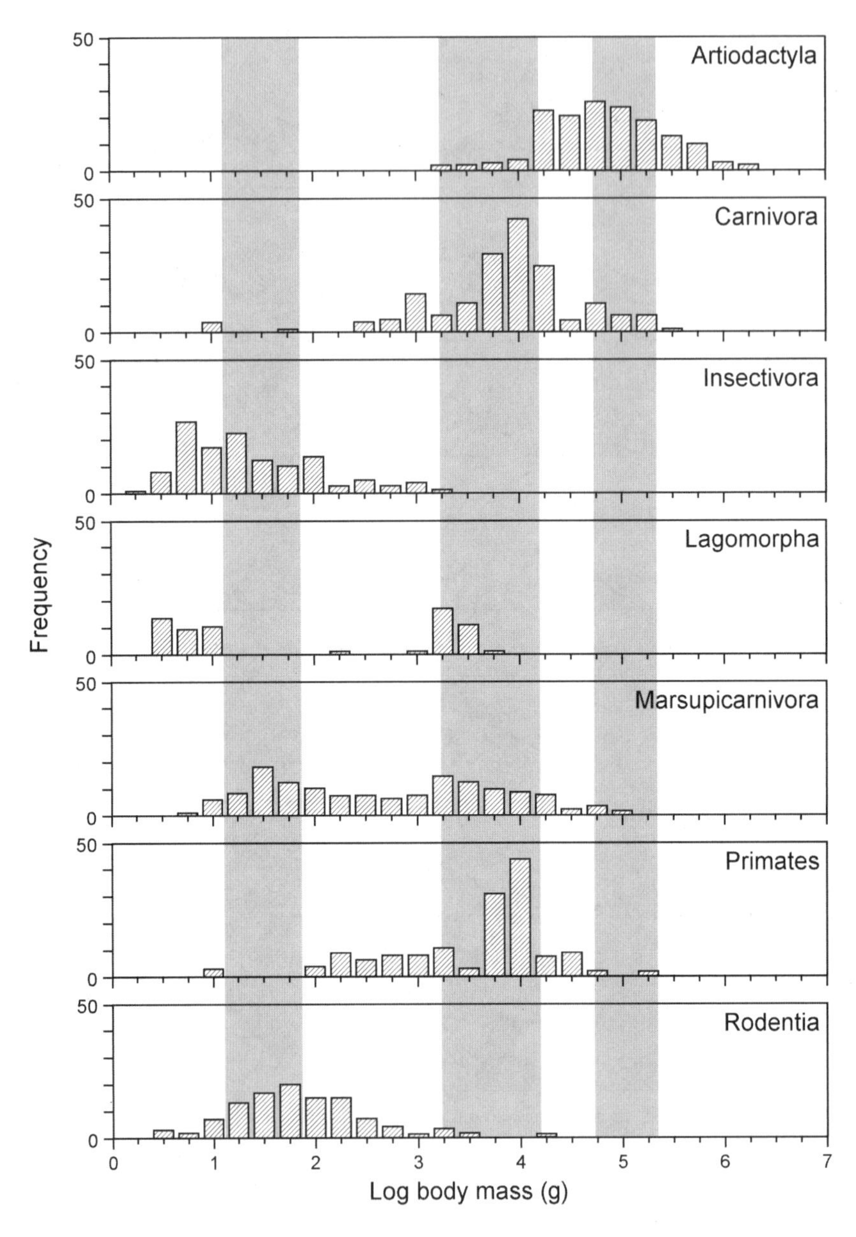

FIGURE 5.3 Body-mass frequency distributions of species in seven different mammal orders created by one hundred random draws of mean adult values (Silva and Downing 1995) for each order. Light-gray zones indicate areas of highest pattern matching (locations of gaps and clumps) between the body-size distributions of different mammal species assemblies found to occur in the same biome. For some fifteen biomes, pattern matching was measured between body-size distributions of groups of four to five species assemblies per biome.

of discontinuous landscape structure (TDH) or from similarities of taxonomy (PBH). I tested these predictions by examining patterns of overlap for taxonomy and for aggregations at two scales: between two entirely different species assemblages in one biome and between groups of different assemblages within a biome for twenty-one different biomes. The smaller-scale case tests all three hypotheses to explain a high degree of pattern overlap between two species assemblages. The larger-scale case tests the generality of pattern overlap across a variety of biomes and explores the degree to which such pattern correspondence might be the product of chance.

METHODS AND DATA In the smaller-scale case, I used the GRI to determine the body-mass structure and Simpson's Index of Species Similarity to determine the taxonomic overlap of mammal species and of genera at two African savanna-with-forest sites (Gabiro in Rwanda and Amboseli in Kenya). In order to determine if body-mass pattern helps predict the scale at which a species exploits the environment, I examined the natural histories and habitat use of species in "corresponding" aggregations, or body-mass aggregations that occupy the same position on the size axis, but that occur in different species assemblages.

In the second, larger-scale case, I tested in a wide range of biomes the consistency with which similar body-size patterns are found between different species assemblages in the same biome. I used the GRI to determine the number of aggregations and their locations on the log body-mass size axis for ninety species assemblages from twenty-one biomes (table 5.3). I sorted all body-mass distributions into groups of species assemblages found in the same biome (biome groups). I examined the degree to which size clusters (aggregations) and discontinuities occur at the same locations on the size axis in the body-mass size distributions of different assemblages. I divided the size axis into "bins" of fixed length (0.01 $\log_{10}$ body mass) to create a vector of decimal values to quantify the degree to which body-size pattern is shared between assemblages in a biome group. For each bin, I assigned a decimal fraction calculated as the number of assemblages with an aggregation at that location on the size axis divided by the total number of assemblages in the biome group. In each vector, the decimal fraction values varied from 0 (no aggregation in any assemblage at this bin location) to 1 (all assemblages had an aggregation at this bin location). Each assemblage body-mass distribution had a unique range of body-size values and therefore a unique vector length. I made numerical comparisons between data sets in a biome only along those portions of the size axis in which mammal species' body sizes existed in all sets.

I consider conspicuous similarities in body-size pattern to occur when 80% or more of the assemblages in a biome shared a discontinuity (Prominent Gap Pattern) or an aggregation (Prominent Lump Pattern) at the same log body-mass bin. I indexed this high degree of shared pattern for each biome assembly by recording the percentage of the vector occupied by either prominent discontinuity or aggregation (Index of Prominent Lump Pattern [IPLP]) for each biome group. By regressing IPLP values for each biome with mean values of Simpson's Index of Species Similarity for all assemblages within a biome, I tested the PBH prediction that taxonomic similarity creates high values of lump pattern similarity between species assemblages.

I tested the degree to which chance associations might cause IPLP values observed for actual assemblages by measuring IPLP values for mock groups of mammal assemblages in which discontinuities were placed at random in the body-size distributions of mammal assemblages. Each mock size distribution had the same range of body sizes, number of clumps, and size range within each clump as I found in an actual mammal species assemblage; only the locations of the clumps were randomized. For each observed biome group, I created one hundred mock groups, and I measured mean and standard error statistics of prominent discontinuity and aggregation patterns.

RESULTS I found strong similarities in the body-mass size patterns of two mammal assemblages living in the same biome, African woodland savanna (fig. 5.4). Not only do the discontinuities and aggregations occur at very similar locations, but the landscape texture used by different species in each ecosystem corresponds best with the aggregation (size class) in which they occur. The discrete discontinuities that separate aggregations in body-size distributions are associated with qualitative differences in landscape structure used by

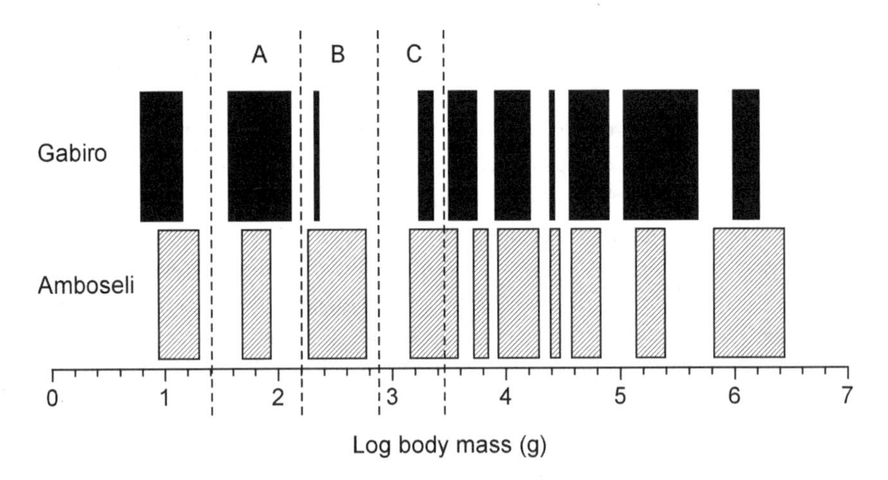

FIGURE 5.4 Discontinuities in body-mass patterns of two African savanna (Gabiro and Amboseli) forest mammal communities identify the distinctions in the scale and texture of landscape use by the species in the body-size zones A, B, and C. With jumps in body size between each zone, landscape textural use grades from very fine (A: burrows under sandy or wet soil, herbaceous mats) to fine (B: extensive ground cover, hollows, holes, crevices) to coarse (C: fringes of water, forest, and open areas).

an animal, with larger-size classes using coarser landscape features in the ecosystem. The PBH's prediction that such clump-pattern correspondence results from taxonomic overlap is not supported at the species level (Simpson's Index = 0) and has only moderate support at the genera level (Simpson's Index = 0.5).

The location and number of aggregations from different data sets are similar when those body-size data sets are taken from sites in the same biome. The PBH's prediction that body-mass pattern correspondence largely reflects taxonomic overlap is not supported at the species level. For twenty-one different biomes, mean values for Sorensen's Index of Species Overlap range from zero to 0.71 (standard deviation = 0.23). However, I found no relationship (fig. 5.5) when I correlated species similarity using the IPLP. Such lump-pattern correspondence (fig. 5.6) is not the product of random events. Taken individually and in the aggregate, indices of prominent lump and gap pattern (IPLP) were higher in actual biomes than in mock biomes (fig. 5.7).

At scales ranging from landscapes to biomes, body-size patterns of assemblages of species may reflect multiple processes operating over different ranges of scale (Allen et al. 2006). Therefore, any single, exclusive explanation for discontinuous body-size patterns should be treated with caution. In this analysis, I found

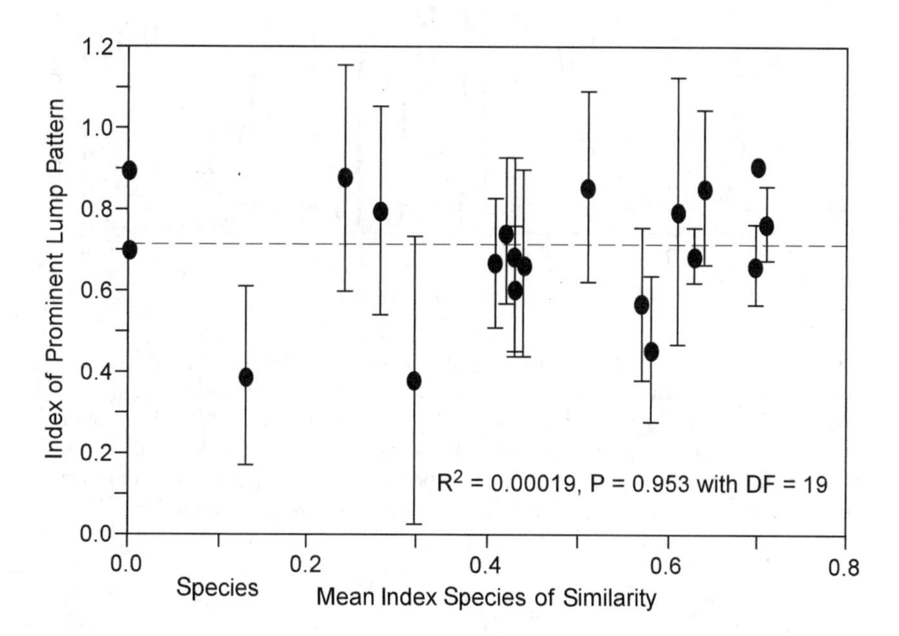

FIGURE 5.5 Correlation between indices of species similarity and body-mass pattern as shared between different mammal species assemblages. Species similarity is indexed by mean values of Sorensen's Index of Species Similarity.

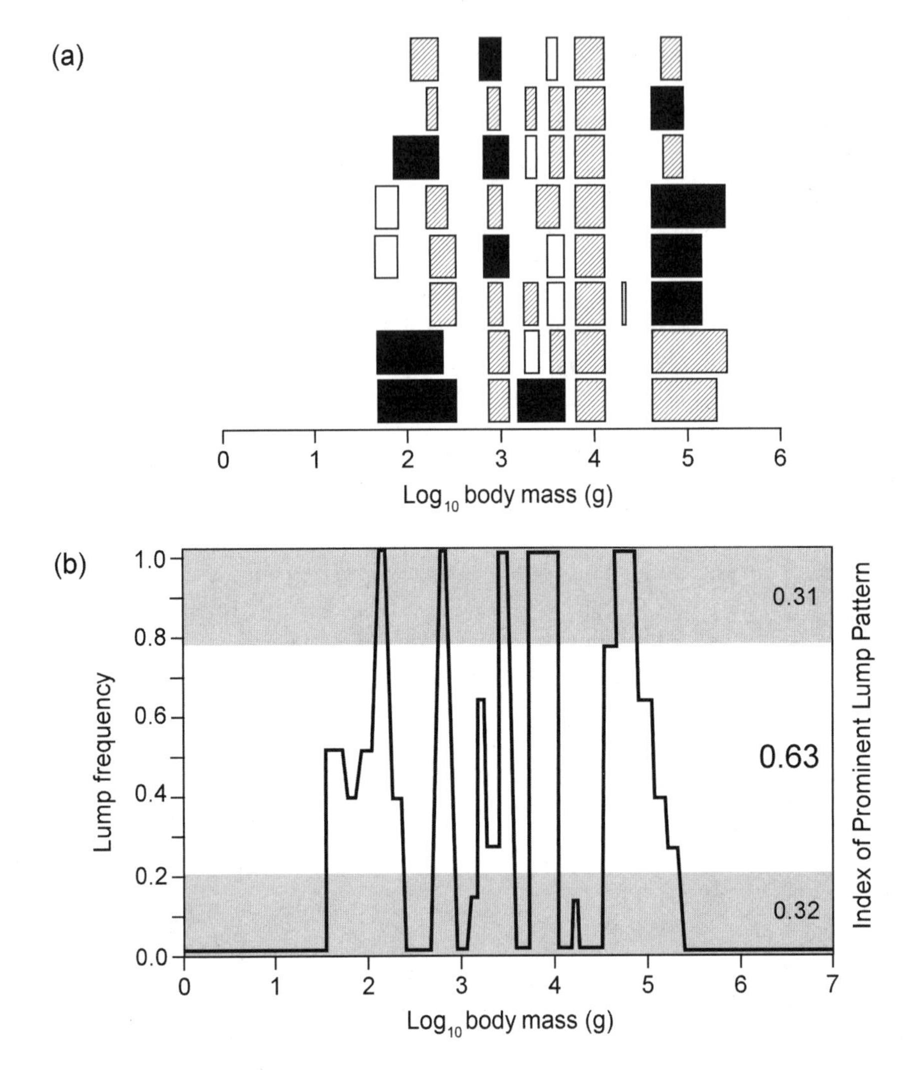

FIGURE 5.6 Patterns of body-mass aggregation common to eight temperate desert mammal species assemblages as shown visually *(a)* in stacked formation and *(b)* as indicated by a vector of aggregation (clump) frequency depicted by a black line. Shaded areas denote high-frequency values (more than 80% of the assemblages in the group) at which body-size pattern is considered prominent. The percentage of the vector represented by prominent gaps (0.32) and clumps (0.31) is summed by the Index of Prominent Lump Pattern (0.63).

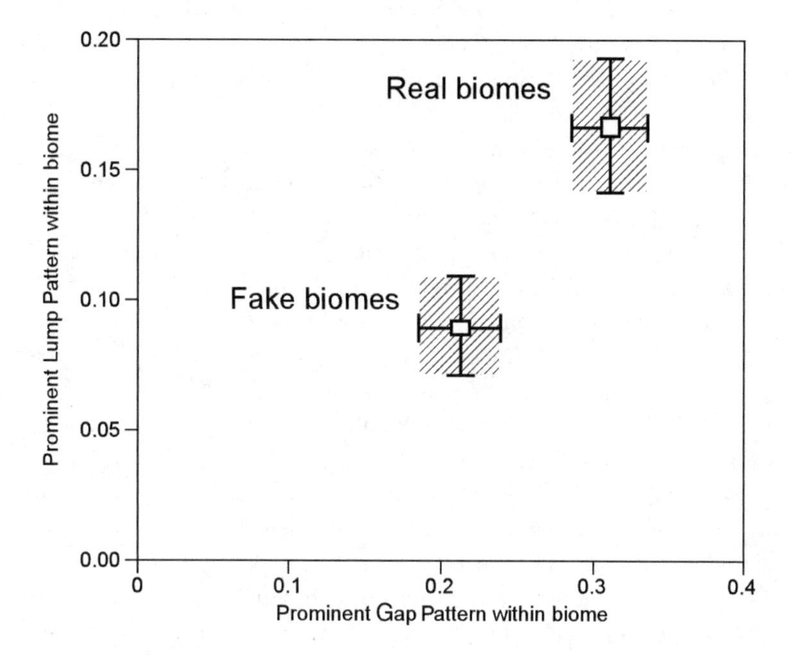

FIGURE 5.7 Mean and standard error statistics for the Index of Prominent Lump Pattern (common locations of clumps and gaps shared by a supermajority of all sites) within biomes. Values were calculated for observed (real) groups of landscape-level animal lump patterns and for groups of clump patterns that are simulations of randomized clump patterns. The values index "strong" or prominent pattern—the proportion of the animal body-mass size range occupied by gaps (x axis) or clumps (y axis) in 80% or more of the landscape data sets within the group.

no support for the Sampling Error Hypothesis or the TTH and support for the TDH. Despite the paucity of concrete support for the PBH (Urtier and Core-Taxa hypotheses), the intriguing suggestions indicated by my initial probes make it premature to abandon these hypotheses. Here, I review the evidence for each hypothesis and then discuss how phylogeny and landscape pattern may make different though complimentary contributions to shape body-size patterns in assemblages of mammal species.

This study extends the range of ecosystem types in which discontinuities are observed in the body-size distributions of terrestrial mammals to the point of turning the original question on its head: Where do we *not* find discontinuous size distributions? More important, it exposes the landscape-based TDH to a strong test. If body-size distributions reflect mammal interactions with landscape structure, then certain body-size patterns would be consistently associated with particular landscape structures. I found that the location of aggregations and gaps in body-size distributions do tend to line up, but only when the size distributions come from species assemblages living in similar landscapes (i.e., within the same biome type). This consistency of body-size pattern correlation is greater than expected by chance. Using different methods to detect discontinuities, Siemann and Brown (1999) did not find any similarities in pattern between the body-size distributions of sites in the same biome, but they suggested that any possible pattern similarities would be explained by taxonomic overlap. Species similarity between different sites in the same biome ranged from zero to 0.71, but I found no relationship between indices of species similarity and indices of similarity in body-size pattern. Furthermore, two mammal assemblages with no species overlap that inhabit the same biome in Africa showed striking similarities in the locations of discontinuities and aggregations. In this case, it was not taxonomic similarity, but similarities in the scale of landscape architecture exploited that best predicted the locations of discontinuities and aggregations. Although evidence supports the TDH, it is quite reasonable to expect at the macroscale of landscapes that multiple processes (trophic or phylogenetic or both) influence body-size patterns of assemblages that exploit landscape textures that range in scale from centimeters to kilometers.

The possibility that predator-prey relationships require substantial size differences has been suggested as the basis for multimodal size distributions in aquatic environments (Kerr 1974). Real size differences consistently characterize predator-prey relations on land as well. Terrestrial predators tend to be larger than their prey (Vézina 1985; Warren and Lawton 1987; Cohen et al. 1993), but this relationship, although generally positively correlated, is not linear. The predator-prey size ratio tends to decrease with trophic height (Jonsson and Ebenman 1998).

Predator-prey size differences are a general trend, but it is not clear that these differences result in discontinuities in body-size distributions. Holling (1992) demonstrated close similarities in body-mass pattern between two broad trophic classes of boreal prairie mammals (carnivores versus herbivores/omnivores). If trophic relationships established size differences and were the basis of discontinuities in body-mass distributions, then the two trophic groups should be distinguishable by very different body-mass patterns. Mammal species' size distributions of either trophic class do not aggregate in any size range; they generally span the entire size axis in the six terrestrial biomes I investigated. More specifically, chi-square tests show that any body-size clustering of either mammal trophic group has little relation to body-mass aggregations. Rather than aggregations predominately composed of carnivores or of herbivores/omnivores, I found both trophic groups in practically every body-mass aggregation. This conclusion supports Holling's (1992) finding in boreal prairie mammals that the distinct similarities in the body-mass patterns of either trophic group suggest that clumpiness and discontinuities arise because both carnivores and herbivores/omnivores respond to the same processes. The coexistence of both trophic groups in each aggregation also suggests the same diversity of function within a scale range and redundancy of function across scale ranges proposed by Peterson, Allen, and Holling (1998). These patterns lead me to conclude that size differences inherent in trophic relations do little to explain discontinuities in terrestrial mammal body-size distributions.

Shared ancestries and different time spans of adaptive radiation give a rich complexity to the phylogeny of mammal orders (Gardezi and da Silva 1999). However, the radiation of body sizes for each order probably started from a single ancestor, and gaps in a size distribution might occur if the size spectrum has not been saturated by all these separate radiations (Urtier Hypothesis). At the scale of individual aggregations in a landscape-study site, I did not find a correspondence between patterns of body-size aggregations and any pattern of clustering created by the radiations that gave us present-day mammal orders. In a majority of cases, patterns of membership in taxonomic orders and in body-size aggregations appear randomly related within mammal species assemblages. Only 8% of the tests indicated nonrandom relations between aggregations and membership in two taxonomic orders.

Although these tests offer little or no support for the Core-Taxa or Urtier hypotheses, phylogeny may contribute in some broader way to discontinuous body-mass patterns (Marquet and Cofré 1999). There are consistent similarities in body-mass patterns among different mammal assemblages (fig. 5.3). The greatest density (modes) of body size for five mammal orders (Artiodactyla, Insectivora, Carnivora, Marsupicarnivora, and Rodentia) is identified across those zones where body-size aggregations are most often found to coincide in their location on the size axis. The body-size distributions of each order do not span the entire size axis and so appear to illustrate the Urtier and the Core-Taxa hypotheses' contention that each order is the product of limited radiation in size from a common ancestor, with the possible exception of the Marsupicarnivora. However, although these modes in body-size density suggest why species from certain orders tend to be found in certain body-size aggregations, they offer no explanation for the discontinuities that subdivide the distributions into separate body-size aggregations. Landscape structure is the only other characteristic shared across these different study sites within a biome, so body-size discontinuities that separate aggregations are most closely associated with patterns of landscape structure.

I propose the Landscape Taxa Hypothesis to reconcile the separate bodies of evidence supporting the TDH at the scale of aggregations, on the one hand, and the Urtier and Core-Taxa hypotheses at the scale of body-size distributions for taxonomic order, on the other. The Landscape Taxa Hypothesis posits that lumpiness (discontinuities in body-size distributions) and the presence of certain species in the resulting body-size aggregations reflect a two-stage process that organizes mammal species assemblages (fig. 5.8) at two scales. The influence of phylogeny is evident at scales larger than landscapes or ecosystems. The regional pool of mammal species creates a body-size distribution whose order-specific modes

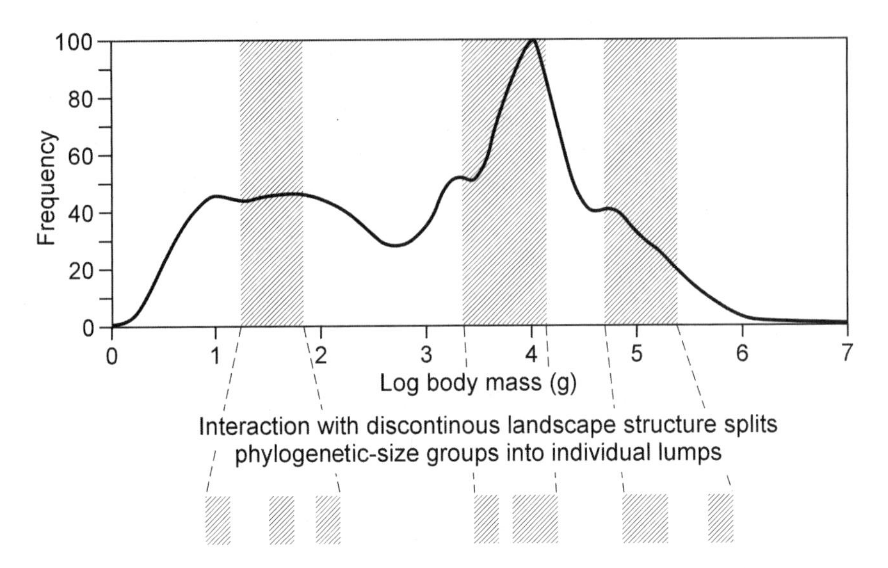

FIGURE 5.8 The Landscape Taxa Hypothesis, a synthetic model integrating interactions between regional phylogeny-based body-size patterns and local discontinuities in landscape structure (indicating animal community lump structure).

constitute a macroscale template from which different species are selected in the organization of species assemblages in particular landscapes. The influence of landscape structure is evident at the scale of aggregations, or the clusters of animal species with similar mass that use a particular texture or scale range in the landscape. Discontinuities in the phylogenetically derived size template appear as animals interact with discontinuous, scale-specific landscape structure and either remain and reproduce or depart from the landscape or die. I ascribe gaps in body-size patterns of mammal assemblages to discontinuities in the habitat architecture because landscape structure is the only variable that is consistently associated with similarities in the locations of aggregations and gaps. The discontinuities and aggregations in different body-size patterns tend to coincide if the two separate communities occur in very similar landscapes, and they do so far more frequently than expected by chance. Mammal body-size distributions enrich the picture of landscapes as complex adaptive systems by indexing the operation of two processes at different scales (phylogeny and animal interaction with habitat architecture) in the dynamic organization of species assemblages.

6

BIOPHYSICAL DISCONTINUITIES IN THE EVERGLADES ECOSYSTEM

Lance H. Gunderson

UNDERSTANDING HOW ecosystems are structured and function across a wide range of scales is important for both practical and theoretical reasons. The increasing span of human influence on the planet (Vitousek et al. 1997) is but one indicator of how processes at one scale aggregate to create emergent dynamics at larger scales. The resolution of many resource issues is plagued by uncertainty introduced by processes operating across different scales (Ludwig, Hilborn, and Walters 1993; Folke et al. 2002). Indeed, many of the surprises chronicled in histories of ecosystem management are a result of such emergence, or cross-scale, dynamics (Gunderson and Holling 2002; Gunderson and Pritchard 2002). A typology of ecosystem surprise suggests three kinds—local, cross scale, and true novelty—each of which is related to cross-scale dynamics (Gunderson 1999). Attempts at assessing and managing surprising ecosystems are related to theoretical underpinnings that frame resource policies and actions (Gunderson and Pritchard 2002).

Ecologists have used hierarchy theory as an organizational framework for understanding cross-scale structure and dynamics in ecosystems (T. Allen and Starr 1982; T. Allen and Hoekstra 1992; O'Neill et al. 1986). This theoretical base emphasizes a pattern of aggregations (hierarchical levels, or holons, characterized by tight coupling) that are separable across scales and predicts that such levels in ecological hierarchies can be identified. In the 1980s, ecologists applied new techniques and discovered that structural features in ecosystems such as coral reefs (Bradbury and Reichelt 1983; Bradbury, Reichelt, and Green 1984) or lakes (Kent and Wong 1982) did indeed vary as researchers changed the scale of observation. Both groups of authors attribute these breaks to the fixed-scale range of key processes that influence these structures.

Holling (1986, 1992) proposed that ecosystem processes should interact across a range of scales to produce structural features of aggregations and discontinuities. This proposal came from a comparison of the dynamics of multiple ecosystems, wherein Holling (1986) concluded that a system's essential behavior can be traced to a small set of variables. Moreover, the characteristic rates of each of these critical or keystone variables differ from each other by as much as an order of magnitude, so that the time constants are discontinuous in distribution or occur as a small number of nested cycles. This conclusion led to the hypothesis that the system should be structured in such a way that the keystone variables entrain other variables. The entrainment should occur in both spatial and temporal dimensions. Spatial features should therefore express the creation of structural features that exhibit distinct gaps and clumps and temporal processes that show distinct periodicities. An overt manifestation of a clumpy, discontinuous world should be expressed by attributes of the animals that live in these systems. This hypothesis was challenged by a series of tests using data from three biomes (Holling 1992). The tests using the adult body mass of birds and mammals from the boreal forests, prairies, and pelagic ecosystems indicated the presence of discrete gaps that defined groups (Holling 1992). All alternative hypotheses using developmental, historic, or trophic explanations for the groupings were invalidated, leaving only the strong inference that ecosystems (abiotic and biotic components) were similarly organized into scale-invariant structural features that are separated by distinct breaks. Subsequent comparisons of pre- and postextinction faunal assemblages (Lambert and Holling 1998) and comparisons of rare, threatened, and nonnative components of faunal assemblages (Allen, Forys, and Holling 1999; Allen 2006) supported the Textural Discontinuity Hypothesis. That is, the discontinuous distribution of animal body masses echoes the cross-scale structure of the biophysical environment that supports those animals.

The theory base presented here suggests that keystone ecological processes are organized in discrete groupings across space or time scales or both. This chapter separates the space and time dimensions and proposes two hypotheses: (1) that discontinuous patterns of keystone variables should be manifest as breaks in ecosystem structural features, and (2) that clustering or grouping of keystone processes should be exhibited as a few dominant frequencies in temporal processes. For this study, these hypotheses were put at risk using a series of data from the Everglades ecosystem. The Everglades ecosystem is structured by a small number of processes in which hydrology and vegetation are one set of keystone variables. Over a range of scales, patterns produced by vegetative and hydrologic processes should have a few characteristic domains in space or time that are separated by discontinuities. Within the time domain, if the two proposed hypotheses are true, a few dominant frequencies or cycles should emerge in the hydrologic processes of rainfall, water level, flow, and evaporation. Within the spatial domain, a few groupings of object size (such as vegetation patches) or texture should emerge that correspond to levels in a spatial hierarchical structure. Other ecosystem-level processes, such as topography and fire, should exhibit similar patterns.

METHODS

A set of methods was applied to a diversity of data sets, rather than a single method being applied to a single data set. All the methods covered a wide range of scales (either in space or time) and involved techniques to uncover pattern by changing the level of resolution and window size within a data set. The level of resolution is defined as the smallest unit of homogeneity (such as a step length in one-dimensional spatial data [transect], a picture element or pixel in two-dimensional spatial data [map], or time step in temporal data [time series]). The window is the extent of data coverage, such as the length of an elevation transect, frame of a map, or length of a time series. Three primary techniques were used: identifying gaps in distributions, identifying breaks in fractal dimensions, and Fourier analyses. These techniques are described later.

In assessing distributional gaps, all data were rank ordered. The Silverman Difference Index (Silverman 1986) and the Holling Body-Mass Difference Index (Holling 1992) were calculated for each set of consecutively ordered observations. Discontinuities within rank-order distributions were determined by the pattern of difference indices wherein a significantly large index value followed by a small value is indicative of a gap followed by a clump (data aggregation). The Silverman Index has a known distribution, and hence significance of gaps can be assessed by its use.

Fractal dimensions were assessed for one- and two-dimensional structural data by varying window and pixel sizes. Changes in the scaling or fractal dimension indicate changes in structural properties and can be detected by changes in the linearity of relationship between log of total measure and log of step length (Mandelbrot 1983; Morse, Lawton, and Dodson 1985; Feder 1988; Milne 1988).

A fast Fourier transformation was used to analyze for spectral densities in time-series data. Dominant frequencies were determined as statistically significant peaks in spectral plots.

Two types of data were used in cross-scale analyses: those that represented spatial features and those from time series (table 6.1). The spatial data are from key structural features of the Everglades. Some of the temporal data sets include measures of key aspects of the hydrologic regime, such as rainfall, stage, flow, and evaporation.

Two data sets were analyzed to test for changes in spatial pattern across scale ranges. Two critical features in the Everglades system are the soil-surface topography and the vegetation. Ground-elevation data along transects were analyzed using one-dimensional fractal methods and Fourier techniques. Patterns from vegetation maps were analyzed by using two-dimensional fractal methods and by looking for discontinuities in patch-size distributions. Vegetation maps were made for windows of 160 m, 1,600 m, and 16 km, with resolutions (pixel sizes) of 0.2 m, 1.0 m, and 10 m, respectively. Sawgrass, wet prairie, and tree island

TABLE 6.1 Summary of Data Sets Used in Cross-Scale Analyses of the Everglades Ecosystem

VARIABLE	SPACE/ TIME	ANALYSIS	GRAIN	UNIT	WINDOW	UNIT	REPLICATES
VEGETATION							
SAWGRASS	Space	Fractal	0.5	m	150	m	12
			2.0	m	1,600	m	25
			10.0	m	16	km	2
TREE ISLAND	Space	Fractal	0.5	m	150	m	2
			2.0	m	1,600	m	25
			10.0	m	16	km	2
WET PRAIRIE	Space	Fractal	0.5	m	150	m	9
			2.0	m	1,600	m	25
			10.0	m	16	km	2
SAWGRASS	Space	Rank Order/	2.0	m	1,600	m	25
		Gap	10.0	m	16	km	2
TREE ISLAND	Space	Rank Order/	2.0	m	1,600	m	25
		Gap	10.0	m	16	km	2
WET PRAIRIE	Space	Rank Order/	2.0	m	1,600	m	25
		Gap	10.0	m	16	km	2
TOPOGRAPHY	Space	Fourier	100.0	m	32	km	5
		Fractal	100.0	m	32	km	11
WATER FLOW	Time	Fourier	1.0	mo	44	yr	1
		Rank Order	1.0	mo	44	yr	1
RAINFALL	Time	Fourier	1.0	yr	44	yr	2
			1.0	mo	39	yr	2
			1.0	day	39	yr	2

TABLE 6.1 *continued*

VARIABLE	SPACE/ TIME	ANALYSIS	GRAIN	UNIT	WINDOW	UNIT	REPLICATES
		Rank Order/	1.0	yr	44	yr	2
		Gap	1.0	mo	39	yr	2
			1.0	day	39	yr	2
WATER STAGE	Time	Fourier	1.0	mo	22	yr	4
			1.0	mo	22	yr	4
		Rank Order/	1.0	mo	22	yr	4
		Gap	1.0	mo	22	yr	4
TEMPERA-TURE	Time	Fourier	1.0	mo	22	yr	2
		Fourier	1.0	mo	22	yr	2
PAN EVAPORATION	Time	Fourier	1.0	mo	22	yr	2
		Fourier	1.0	day	22	mo	2

vegetation types were mapped and analyzed because they represent the spatially dominant vegetation features of the Everglades (Gunderson and Loftus 1993; Gunderson 1994).

Five time-series data sets were obtained. The data sets for rainfall, stage, flow, evaporation, and temperature represent keystone processes. All were analyzed using the Fourier technique to test for the presence of dominant-characteristic frequencies.

RESULTS AND DISCUSSION

Analyses of the data sets on the keystone structural features and processes suggest that the discontinuity hypotheses cannot be invalidated. Examination of the structural features across scales (grain and extent) implies the occurrence of breaks that separate regions of self-similarity. As the window of examination is changed, a region of similar patterns abruptly changes to another region that exhibits a different pattern. Only a few frequencies were noted in the temporal fluctuations of key processes in the system. Repeated analyses that varied window and grain of the temporal data sets generated results with only a few resonant frequencies.

These analyses are presented in three sections. The first deals with scales of topographic variation, the second with scaling relationships in vegetation structures, and the third with hydrologic components, including rainfall, water stage, water flow, and evaporation.

TOPOGRAPHY

The topography of the soil surface in the southern Everglades appears to vary at two different spatial scales. The break is defined by a change in the slope (fractal dimension) at a horizontal step length of between 1.0 and 3.2 km (fig. 6.1a). A rolling regression technique was used to determine the breakpoint by indicating which points had higher correlation coefficients. The maximum regression coefficient was obtained using values up to $\log_{10} 3.25$ (fig. 6.1b), indicating a break in the fractal dimension at a step length of about 1.5 km. For smaller step lengths, the topography is reasonably self-similar, with a fractal dimension of approximately 1.6 km. A Fourier analysis did indicate multiple sets of waves, but also emphasized the long or broad wavelengths apparent in the data sets. Whereas the fractal analysis emphasizes the small-scale variation, the Fourier analysis appears to emphasize the broad-scale variation. Hence, at the broad scale, the elevation varies on the order of 0.5 m over horizontal distances of 20 km. The finer-scale variation reflects changes of 20 cm over horizontal distances of less than 1.0 km.

These two regions of scale appear to correlate with the processes that create the observed structure. The fine-scale variations in topography (measured by step lengths of less than 1.5 km) are related to processes that can increase or decrease the soil-surface elevation. Soil formation occurs in the areas of longer hydroperiod because decomposition is slow, resulting in peat accretion. Organic soils can be rapidly oxidized if long-term droughts occur or if the soil burns during severe dry periods. Marl is formed on shorter hydroperiod sites, a result of precipitation of calcium carbonate by blue-green algae (Gleason 1972). Over longer terms, solution of the limestone bedrock can create small-scale depressions (Hoffmeister 1974). The macrotopography is associated with the geologic features of the Everglades (Gleason et al. 1984). The geologic features are associated with broad landscape regions, defined by Davis and colleagues (1994), and are the result of broad-scale processes of ocean currents. Another theory (Petuch 1985, 1987) is that the gross morphometry of the Everglades basin was caused by a meteorite impact during the Eocene.

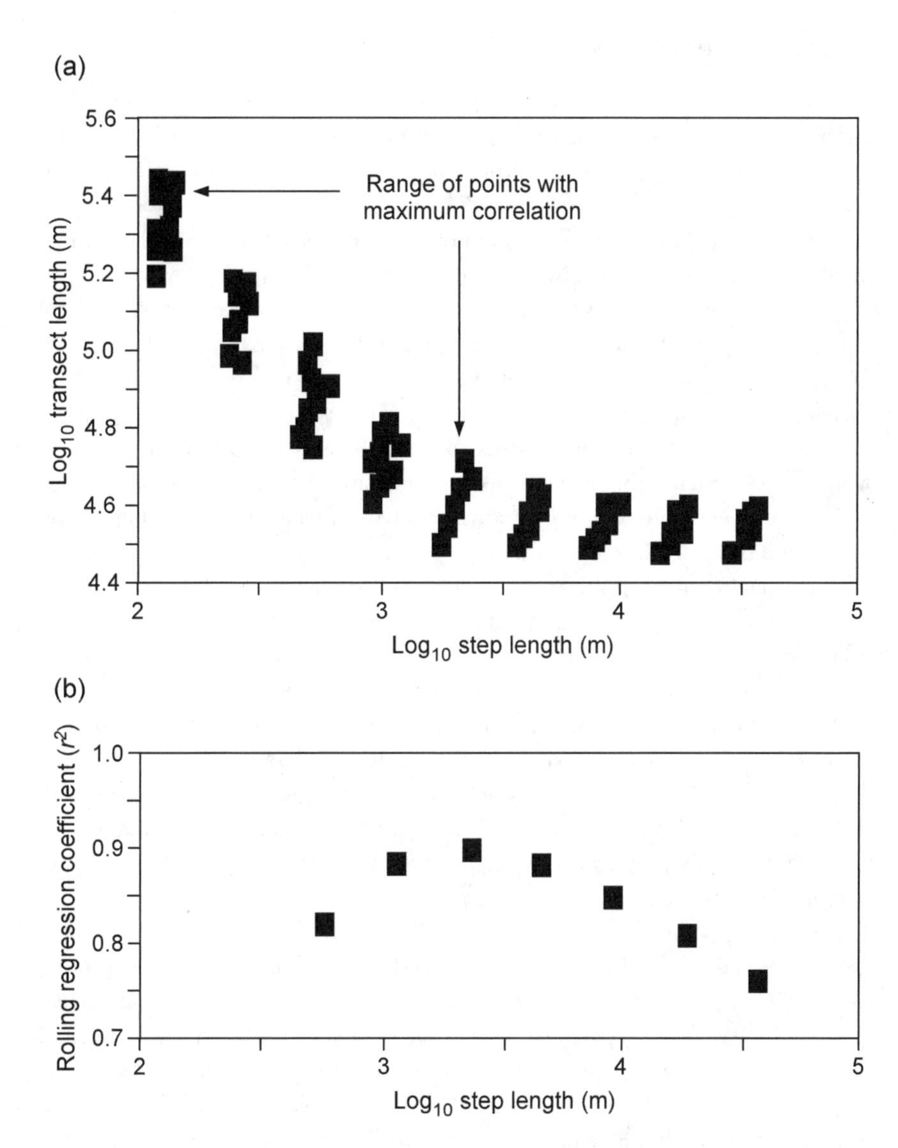

FIGURE 6.1 *(a)* Log-log plot of total transect length as a function of step length. Squares represent different transects of elevation profiles in the southern Everglades. The slope of the linear relationship represents the fractal dimension of the topography. *(b)* Plot of regression coefficients (r^2) versus step length, resulting from rolling regressions. Highest coefficients were obtained at step length of $\log_{10}$ 3.25, indicating a break in the fractal dimension.

VEGETATION PATTERNS

The spatial patterns of three dominant vegetation communities (sawgrass, wet prairie, and tree island) in the Everglades are not scale invariant, which is evident from discontinuities in rank-order plots of patch size and breaks in the fractal dimensions.

A break in scaling relationships was noted for all vegetation types. The scaling relationship was determined by pooling data across three window sizes and averaging total patch size across all sample maps. The fractal dimension appears to change at step lengths of about $\log_{10}$ 2.2 (160 m) for sawgrass (fig. 6.2a), $\log_{10}$ 1.8 (60 m) for wet prairie (fig. 6.2b), and $\log_{10}$ 2.4 (250 m) for tree islands (fig. 6.2c). Because the methods are insensitive and the variation in the estimated parameter is large, a conclusion as to presence of these changes or breaks cannot be reliably made. To determine if the breaks are indicative of a change in underlying process, another technique was used: an analysis of gaps in the distributions of vegetation patch sizes.

The analyses of patch sizes from the 1,600-m window indicate the presence of three clumps in the sawgrass and wet prairie vegetation types and two in the tree island type. Although significant gaps occur at the beginning of all of the distributions in all vegetation types, none of them are counted. Significant gaps, followed by small index values, occur in the distribution of sawgrass patch sizes at two points. The first clump includes all patch sizes less than about $\log_{10}$ 4.8 (63,000 m^2). The second clump is defined by points greater than $\log_{10}$ 4.8 and less than $\log_{10}$ 5.6 (316,000 m^2). Significant gaps in the distribution of wet prairie patch sizes occur at $\log_{10}$ 5.4 and $\log_{10}$ 5.9. Similar gaps occur in the tree island distributions at $\log_{10}$ 4.2.

Both the gap and the fractal analyses indicate similar properties of the vegetation patterns displayed in maps. The gaps in the rank-order distributions seem to occur at patch sizes not very different from those inferred from the fractal analyses. All three vegetation types exhibited a gap or change in the fractal dimension at step lengths between approximately 50 and 300 m (table 6.2). Another break appears at a step length of approximately 1,000 m (table 6.2). Although none of the individual results alone provides irrefutable proof, the results as a whole provide evidence that supports the hypothesis that the system is structured in a series of lumps and gaps.

The observed breaks in vegetation patterns are puzzling. No plausible explanation can suitably account for the patterns. The breaks do not occur in the smallest windows (less than 160 m) and therefore do not represent the edge profiles of communities. The breaks occur in the larger windows and appear around step lengths of 300 m and 1 km. The break at the largest scale may be related to the topographic features, although it occurs at a smaller step length than the change in the topographic fractal dimension. The breaks appear at about the

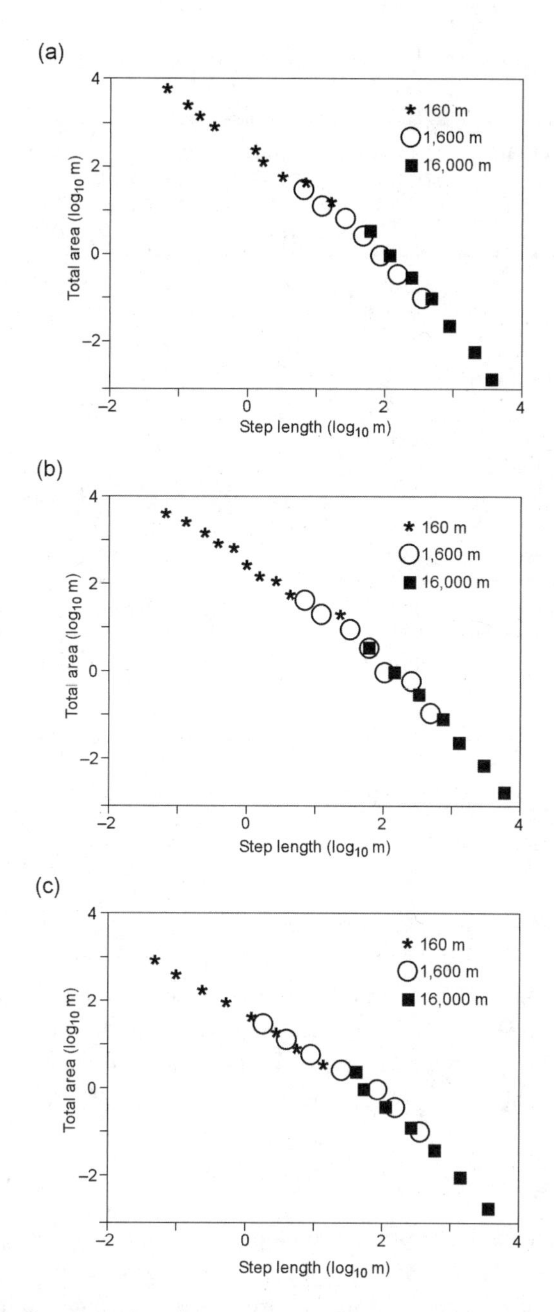

FIGURE 6.2 Log-log plots of patch-size area as a function of pixel size for three vegetation types in the southern Everglades: *(a)* sawgrass vegetation, *(b)* wet prairie vegetation, and *(c)* tree island vegetation. The plots were derived by varying the pixel (tile) size and by measuring total patch size in windows of 160 m, 1,600 m, and 16,000 m. A change in the linearity of the relationship indicates a break in the fractal, or scaling, dimension.

TABLE 6.2 Results of Cross-Scale Analyses Using Spatial Data Sets from the Florida Everglades

SPATIAL DATA SET	ANALYSIS TYPE	WINDOW SIZE (KM)	SMALL BREAK POINT LOG_{10} (M)	SMALL BREAK POINT (M)	LARGE BREAK POINT LOG_{10} (M)	LARGE BREAK POINT (M)
Sawgrass	Fractal	16.0	2.2	158	—	—
Wet Prairie	Fractal	16.0	1.8	63	—	—
Tree Island	Fractal	16.0	2.4	251	—	—
Sawgrass	Gap	1.6	2.4	251	2.75	562
Wet Prairie	Gap	1.6	2.5	316	2.95	891
Tree Island	Gap	1.6	2.1	126	2.30	200
Sawgrass	Gap	16.0	—	—	3.00	1,000
Wet Prairie	Gap	16.0	—	—	3.05	1,122
Tree Island	Gap	16.0	—	—	3.05	1,122
Topography	Fractal	30.0	—	—	3.20	1,585
Fire Size	Gap	10.0	1.9	79	3.50	3,162

same scale grain across all three vegetation types, which suggests that a meso-scale process such as fire or water flow might be influencing the pattern rather than some biologic process such as spatial expansion of the community size. In spite of the lack of reasons for the observed patterns, the discontinuities appear to exist in the spatial patterns of vegetation.

HYDROLOGIC VARIABLES

Four component variables of the hydrology in the Everglades were analyzed for periodicities and distributional gaps: rainfall, water level (stage), water flow, and evaporation. Each is described in the next four sections, followed by a section that discusses the integration of all variables.

RAINFALL Rainfall data were analyzed over the same time window (January 1949 through December 1988), using three different grain sizes: daily, monthly, and annual. The monthly and annual values were derived by summing daily rainfall totals over each of these time intervals. Two types of analyses were done on each of the three data sets: a fast Fourier transform to examine for cyclical or sinusoidal patterns and a rank-order/gap analysis to test for entrainment of temporal patterns.

The temporal patterns of daily and monthly rainfall totals suggest a dominant annual cycle (fig. 6.3a) characterized by a summer wet season and a winter dry season (Hela 1952; Thomas 1970; MacVicar and Lin 1984). Other significant cycling periods of three and four months were also apparent in the data. No long-term

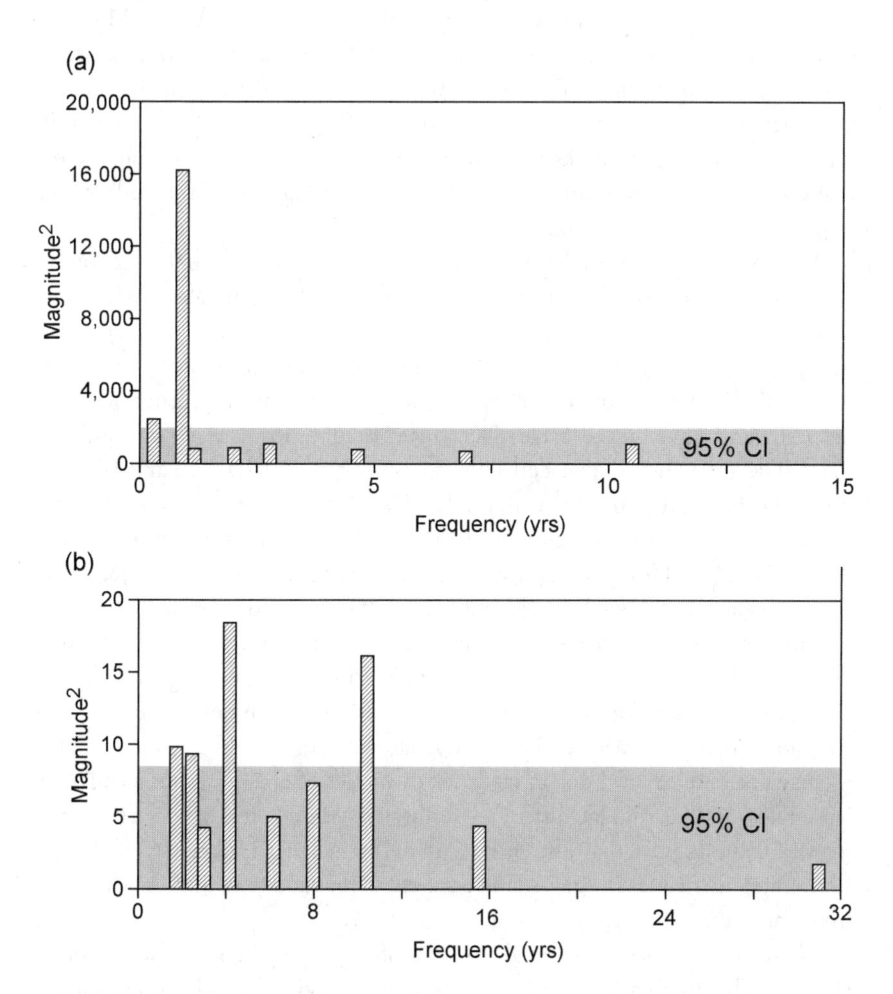

FIGURE 6.3 Spectral density plots from Fourier analyses of *(a)* monthly rainfall data and *(b)* annual rainfall data. Significant peaks in the magnitude of the power spectrum were those that exceeded a 95% confidence interval (CI) around a mean density (gray area).

(multiple-year) cycles were observed from these data sets. The rank-order analyses indicated a few gaps in the distribution, at the high end of the rainfall spectrum. The gaps may be a function of the infrequent occurrence of high rainfall events such as tropical storms and hurricanes.

The annual rainfall totals for the two stations indicate the presence of multi-year cycles (fig. 6.3b). Even though totals varied, cycling frequencies were similar between the sites. Both sites had significant peaks of approximately eleven, five, and three years, similar to the dominant frequencies reported for south Florida by Thomas (1970) and Isaacs (1980).

The analyses of the rainfall data produce different patterns at different grains. The daily and monthly data exhibit only annual and monthly cycles. Multiple-year periodicities are not evident in these data sets. The annual data sets exhibit periodicities of about three, five, and eleven years. It is not clear why the smaller-grain data sets would not show similar patterns. Perhaps the higher variation in the smaller-grain data masks the longer-term patterns. Another explanation is that the summation of rainfall to yearly totals may exaggerate a pattern that is imperceptible in the other data sets.

The dominant cycles in the rainfall data appear to be related to fluctuations in the spatial and temporal domains of the processes that generate rainfall. The dominant cycle evident in both the daily and monthly data is the annual cycle, with peak rainfall during the summer and lower rainfall during the winter. Summer rainfall is a result mainly of convective thunderstorms associated with the daily sea and land breeze cycle (Hela 1952; Bradley 1972; MacVicar and Lin 1984). The generation of convective thunderstorms is related to the annual variation in heat budget associated with the Earth's orbit. During the fall, winter, and spring months (November through April), rainfall is associated primarily with the passage of cold fronts (Hela 1952; Bradley 1972; MacVicar and Lin 1984). The fronts pass at approximately weekly intervals and do not produce much rainfall.

The three- to four-month cycle evident in the data is less well understood or explained. This cycle is evident as the bimodal summer peaks of rainfall. Some authors (Thomas 1970; MacVicar 1983) attribute the summer depression to a combination of two processes. During the late summer months, convective activity may decrease or level due to feedback dynamics of changing surface albedo, lapse rates, and heat budget after the freshwater system is full of water (Gannon 1978). The latter peak may also be a result of the increased frequency of tropical storms and hurricanes in August and September that add to rainfall totals during these months (Gentry 1984).

There is a paradox in the analyses of rainfall data as the grain or resolution changes. The dominant cycles in both the daily and the monthly grain data sets are the annual and monthly frequencies. No multiple-year cycle emerges from analyses using these data sets. The interannual or multiple-year cycles are significant only in the Fourier analysis of annual rain totals. A number of explanations can be offered for these results. The multiple-year cycles may be in the smaller-resolution data sets, but there are so many variations that the signals are masked by all of the data. Attempts were made to mask seasonal and annual variation, and the interannual patterns were still not significant. Another explanation may be that the multiple-year patterns are an artifact of the totaling process over a year's time in that they tend to exaggerate patterns. A number of authors (Thomas 1970; Isaacs 1980) report multiple-year cycling in annual rainfall. Annual variations have been attributed to the degree of tropical storm activity (MacVicar 1983). Other phenomena such as the El Niño Southern Oscillation (ENSO) contribute to increased rainfall over the southeastern United States (Rasmussen 1985; Ropelewski and Halpert 1987). The presence of these longer-term, broader-scale processes, such as ENSO and tropical storm frequency, seem to account for the interannual variation in rainfall, but these effects are probably more evident in rainfall totals summed over an annual time period.

WATER FLOW The flow of water through the southern Everglades also exhibits multiple cycles or periodicities. The spectral analysis indicated the largest frequencies of one and nine years, with minor frequencies at three, five, and twenty-two years (fig. 6.4a). The effects of water management are included in this analysis. Little or no flow occurred during the construction of the water-delivery structures in the early 1960s. In the period of regularity in the 1970s through the early 1980s, the minimum flow regime was in effect, which probably amplified the annual cycle pattern.

The large variation in monthly flow creates discontinuities and aggregations in the rank-order distribution. The aggregations are apparent as steps or risers in the rank-order plot. At least five significant (greater than 95% confidence level) gaps occur at flows of 0.8, 1.5, 2.3, 3.2, and 3.8 by 108 m^3 per month. The gaps can be used to define breaks among flow groupings that correlate with the periodicities listed earlier. Flows greater than 3 by 108 m^3 per month occur on the long-term frequencies (twenty-two years plus). The 3.2-by-108-m^3 per month break seems to correlate with the nine-year return interval. The 2.3-by-108-m^3 per month break is roughly observed in the five-year cycle, the 1.5-by-108-m^3 per month break is observed in the three-year cycle, and the smallest break seems to correlate with an annual cycle. These correlations are approximate; high flows do not occur every nine to ten years. The data indicate distinct periodicities, with dramatic annual and decadal cycles, that appear to correlate with distinct volumetric groupings.

WATER LEVELS For the water levels, the stage data from all four stations in the southern Everglades exhibit patterns of multiple cycles when analyzed using daily and monthly time steps over the twenty-two-year window. The spectral analysis of the monthly data shows the presence of three cycles at all four sites

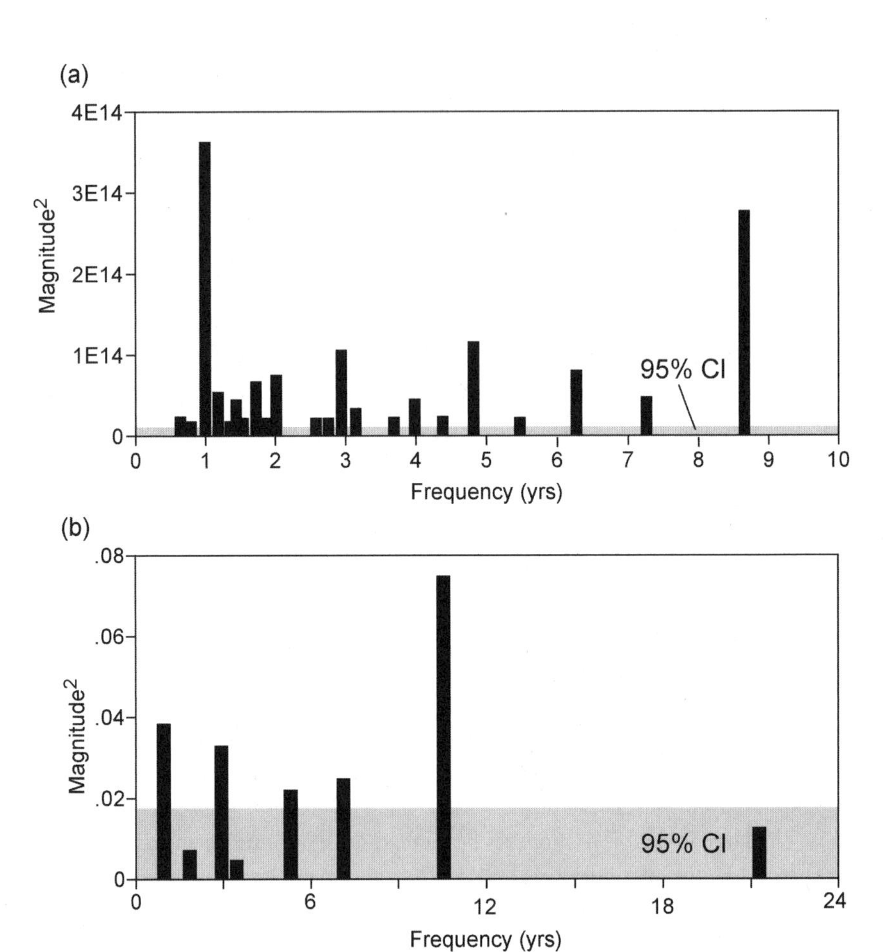

FIGURE 6.4 Spectral density plots from Fourier analyses of *(a)* monthly water flow and *(b)* monthly water-level data. Significant peaks in the magnitude of the power spectrum were those that exceeded a 95% confidence interval (CI) around a mean density (gray area).

(fig. 6.4b). The strongest cycle from the monthly data is of about eleven years, with other peaks at frequencies of one and three years. The daily data also exhibit three peaks in the Fourier analysis, but at cycles of one year, three years, and about seven years.

The rank-order analyses of stage indicated few discontinuities in the daily data; however, a few gaps did appear in the monthly data sets. No gaps were measured in the daily data sets due to the large number of observations. With more than thirteen thousand observations in each of the daily data sets, the only differences between sequentially ranked observations were those associated with the precision of the measurement of water level. Gaps appeared at the low and high ends of the distributions in the monthly data sets. These gaps are probably artifacts of the averaging process, created by the removal of extreme daily data points during a month.

As with the rainfall data, the dominant frequency in the stage data is the annual cycle, although the presence of a multiyear cycle appears in both the daily and monthly stage data. The results of the analyses to determine if the annual and multiple-year frequencies occur in other patterns of surface water are presented next.

PAN EVAPORATION AND TEMPERATURE Pan evaporation rates vary at multiple cycles. The most significant periodicity in both data sets was the annual cycle. Significant multiyear periods of eleven and about five years were also observed in both data sets, although the peaks were not as significant as for the annual cycle. A cycle of about five months was also present.

Minimum and maximum temperatures at both sites exhibited a regular, annual cyclical pattern. The data tend to fluctuate more in the minimum temperatures than in the maximum temperatures. The dominant cycle appears to be the annual seasonal pattern. None of the other sites had a discernable multiyear pattern.

HYDROLOGIC PATTERNS

A paradox arises from the previous analyses because the dominant input cycle is an annual one, whereas the dominant water level and flow cycles are longer term—on the order of a decade. The dominant frequency of the stage and flow data is a multiyear (between eight and eleven) periodicity. This result was surprising, given that it is almost an order of magnitude more than the periodicity of the dominant input signal. A number of alternative explanations may account for this result. An investigation that basin morphometry might be creating the longer-term pattern proved negative (Gunderson 1992). The paradox can probably be explained by the fact that the long-term fluctuations in the stage and flow data in the Everglades may not be solely due to fluctuations in the input (rainfall), but can be caused by longer-term cycles evident in the patterns of outflow resulting primarily from evaporation (Gunderson 1992).

The multiple-year cycles evident in the stage and flow data appear to be a result of variations of both input and output operating at different time intervals. One way of depicting these cross-scale interactions is the use of spatiotemporal diagrams as demonstrated by W.C. Clark (1985) and Holling (1992). The diagrams are logarithmic scales over space (abscissa) and time (ordinant) that cover ranges of about six orders of magnitude. Entities within the diagram are defined by grain and window; in spatial terms, *grain* is the smallest resolution that defines the entity, and *window* is the extent. In temporal terms, the grain is the minimum resolution needed to define a pattern, and the window is the longevity of the entity. Breaks in the fractal dimension of spatial patterns can be used to define breaks between entities in the spatial dimension. Similarly, dominant frequencies appear to differentiate temporal entities or levels. Surface water within the Everglades appears to fluctuate on at least three dominant frequencies (daily, annual, and decadal), as shown in figure 6.5a. The daily and annual fluctuations in stage appear as a result of nested fluctuations of land-sea breeze interactions that produce convective thunderstorms on a daily and seasonal basis. The longer-term fluctuations appear to coincide with long wave disturbances influencing evaporation and, to a less-evident degree, with decadal fluctuations in El Niño that vary rainfall. All atmospheric dynamics are lower and to the right in figure 6.5, reflecting the faster dynamics over broad areas, whereas the surface-water dynamics are slower and occur over smaller areas. A similar hierarchy exists in the vegetation structures (fig. 6.5b).

Evidence of breaks in the structural features of the Everglades ecosystem is shown from a summary of analyses of spatial data sets. The topographic data have two scales of variation, defined by a break in the fractal dimension at a step length of about 1.5 km (table 6.2). The changes in vegetation patterns over different scale ranges are suggested by the fractal analyses, but appear at more definite resolution using the gap-detection method. Changes in the fractal dimensions of the vegetation types are suggested at step lengths of 60 to 250 m. Gaps in rank-order distributions of patch sizes appear at a similar range (100–300 m) and again at step lengths of about 1.0 km. Although the measured values of breakpoints differ, all of the data sets appear to exhibit discontinuities in key structural features.

The dominant frequency in the temporal patterns was the annual cycle (table 6.3). Other frequencies were significant in all of the analyzed data sets. Multiple-year patterns of eleven years were dominant in the stage and fire data sets. Secondary frequencies of eight to eleven years were noted in fluctuations of annual rainfall, stage, flow, pan evaporation, and sea-level data sets. Cycles of three to six months were noted in the rainfall and air temperature data sets. Cycles of multiple frequencies were apparent in all of the analyzed data sets, supportive of the hypotheses outlined in the opening of this chapter.

The analysis of existing data sets from the Everglades ecosystem fails to invalidate the hypothesis that spatial patterns exhibit discontinuities and temporal cycles resulting from the effects of a small number of key processes. No one test is singularly convincing. Comparison of results across methods seems to corroborate findings drawn with other methods. Spatial patterns exhibit scale regions of self-similarity separated by distinct breaks. The soil-surface topography appears to vary at two distinct spatial scales; the broad scale is apparently a result of geologic features, and the small scale appears to be related to the processes or organic soil accretion and removal. The vegetation patterns exhibit breaks between

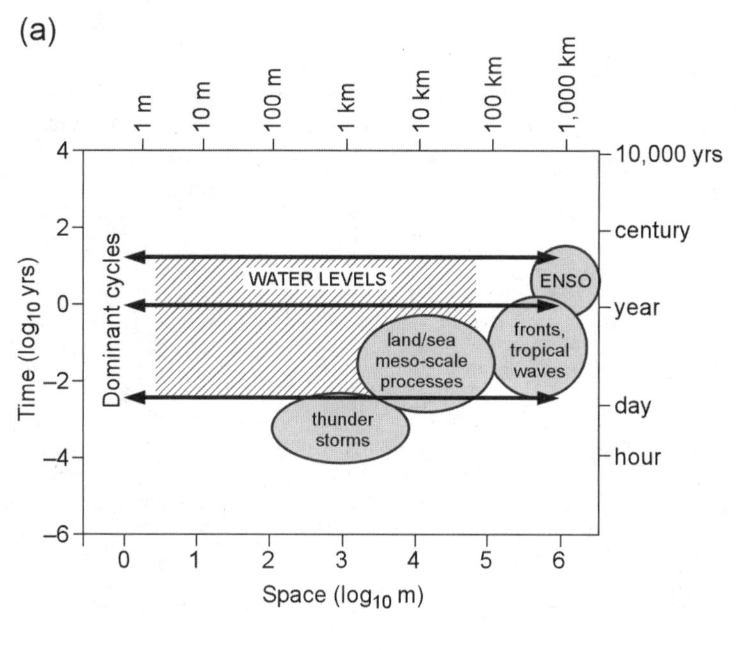

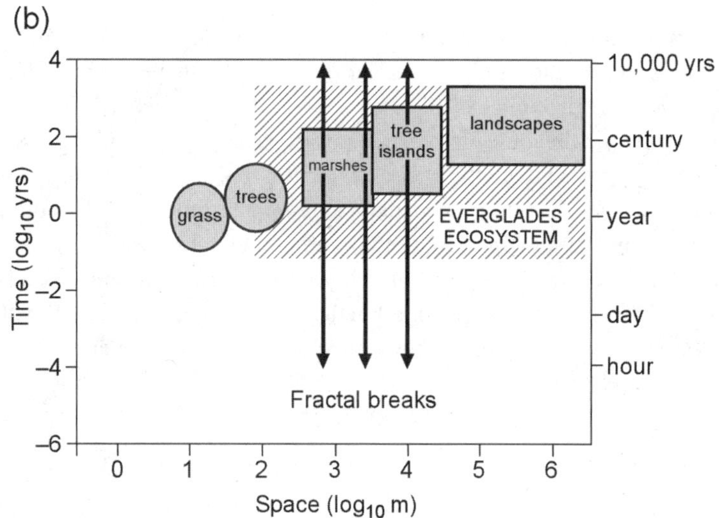

FIGURE 6.5 Hierarchies in space and time for the Everglades ecosystem. *(a)* Hydrologic hierarchy showing scales of dominant frequencies in surface water and sources of atmospheric variation. *(b)* Vegetation hierarchies showing scales of plant species, communities, landscape units, and the Everglades ecosystem as defined by breaks in the fractal dimension.

TABLE 6.3 Summary of Significant Frequencies from Fourier Analyses of Time-Series Data Sets

DATA SET	WINDOW	GRAIN	FREQUENCIES (YEARS)		
			PRIMARY	SECONDARY	TERTIARY
RAINFALL	39 yr	day	1	0.25	0.3
	39 yr	month	1	0.25	0.3
	44 yr	year	6	8.00	11.0
WATER STAGE	22 yr	day	1	7.00	3.0
		month	11	1.00	3.0
WATER FLOW	44 yr	month	1	8.00	22.0
PAN EVAPORATION	22 yr	month	1	11.00	5.0

regions of self-similarity, although the reasons for this result are unclear. Temporal patterns in the stage and flow reflect dominant frequencies in the interplay among the faster dynamics of the atmosphere, the intermediate speeds of the surface water, and the longer-term variations in vegetation, climate, and sea level.

These analyses support the theory that ecosystems are structured around a few keystone variables of mixed spatial and temporal dimensions. Dramatic patterns of discontinuities appear as a result of the interactions within and between hierarchical levels in space and time. This emerging viewpoint of ecosystem structure and dynamics may provide a better basis for understanding the dynamics of the Everglades and hence help to meet multiple management objectives in this unique ecosystem.

7

DISCONTINUITIES IN THE GEOGRAPHICAL RANGE SIZE OF NORTH AMERICAN BIRDS AND BUTTERFLIES

Carla Restrepo and Natalia Arango

ELUCIDATING LARGE-SCALE patterns in plant and animal assemblages is a key step toward understanding ecosystem dynamics and their likely responses to current and future regional and global threats. An attribute that has been widely used in this context is body size because of its well-known relationship with physiological, morphological, and population-level traits (Peters 1983; Schmidt-Nielsen 1984; Niklas 1994; Calder 1996). For example, body size in plant and animal assemblages has been used to characterize energy and nutrient pool sizes as well as fluxes in ecosystems (Kimmel 1983; Wen, Vezina, and Peters 1994; Cyr and Peters 1996; Cyr, Downing, and Peters 1997). In addition, it has been used to understand variation in species diversity (Harris, Piccinin, and van Ryn 1983) and to evaluate the response of ecosystems to disturbance, including climate change (Sprules and Munwar 1986; Jacobs 1999). Less used in this context have been home range and geographical range size, two attributes that have a strong spatial component and that therefore may be informative of processes underlying the distribution of individuals and populations in space (Brown, Stevens, and Kaufman 1996; Maurer and Taper 2002).

The *geographical range* is the basic biogeographical unit and represents the total area over which a species is found (Brown, Stevens, and Kaufman 1996; Gaston and Blackburn 2000). It has been described in terms of its structure and size. Whereas *structure* indicates "how" population demographic attributes are distributed in space (Brown 1984; Villard and Maurer 1996; Brewer and Gaston 2003), *size* indicates the range of abiotic and biotic conditions that a species can tolerate (Gaston and He 2002, and references therein). *Shape,* a third attribute of geographical range, has been postulated to reflect the limitation of ecological factors, including the physical structure of continents (Rapoport 1982; Ruggiero

We are grateful to William Beltrán, Andres Cuervo, Sylvia Heredia, and Rodney Rodríguez for helping put together the bird and butterfly data sets and to Craig Stow and Jennifer Skillen for their valuable comments on early versions of this manuscript. This work was funded by National Science Foundation grant NSF-CREST (HRD no. 0206200).

2001). Thus, the geographical range reflects past and present conditions influencing the large-scale spatial dynamics of plant and animal populations.

Of these three attributes, geographical range size may be particularly informative of processes underlying the origin and maintenance of species diversity. First, the frequency of small and large geographical ranges may suggest conditions favoring speciation and extinction, including such conditions as the expansion and contraction of geographical ranges within given taxa and regions (Gaston and Blackburn 1997; Vilenkin and Chikatunov 1998; Webb and Gaston 2000; Crisp et al. 2001; Jablonski and Roy 2003). Second, the distribution of geographical range size may be used to compare assemblages and to establish whether a similar suite of processes can explain large-scale patterns of species diversity (Gaston 1998; Gaston et al. 1998; Paulay and Meyer 2002). Third, geographical range size is an important criterion in identifying species' vulnerability to large-scale disturbances, such as those resulting from human activities (Terborgh and Winter 1983; Kunin and Gaston 1993; Mace 1994; Angermeir 1995; Arita et al. 1997; Jones, Purvis, and Gittleman 2003). Finally, understanding patterns in the distribution of geographical range size may help design plans for the long-term preservation of the evolutionary and biogeographical processes that underlie the origin of species diversity (de Klerk et al. 2002; Hughes, Bellwood, and Connolly 2002; Jansson 2003).

Geographical range size has been expressed in several ways, depending on whether range maps, information on species' latitudinal/elevational limits, presence/absence, and abundance for a given region are available (Gaston 1994; Brown, Stevens, and Kaufman 1996; Gaston et al. 1996; Quinn, Gaston, and Arnold 1996). In general, species differ widely in the size of their geographical range such that a large number of species are narrowly distributed, whereas a small number are widely distributed (Gaston 1990, 1998; Brown, Stevens, and Kaufman1996; but see Hughes, Bellwood, and Connolly 2002). In more quantitative terms, these right-skewed distributions of range size have been shown to resemble unimodal, continuous, log-normal distributions (Gaston 1996; Gaston and Blackburn 1997), and several explanations for such patterns have been proposed (for a summary, see Gaston and Blackburn 2000; McGeoch and Gaston 2002). Surprisingly, most of the explanations are based on processes operating over ecological or short-term scales that do not necessarily match the long-term scales associated with evolutionary and biogeographical processes involved in the origin, expansion, and extinction of species.

Alternatively, one may ask whether the distribution of geographical range sizes exhibits patterns of discontinuity or multimodality, as has been shown for body size (Holling 1992). A multimodal distribution in range size suggests the presence of discontinuities, and it follows from Holling's Textural Discontinuity

Hypothesis (TDH) that modes in range size should be associated with attributes that are discontinuous in space and time and that are known to have influenced evolutionary and biogeographical processes. The distinctive nature of landforms and the characteristic rates of processes that give rise to them (Brundsen 1996) offer a natural way to evaluate the TDH (Holling 1992) in a biogeographical context. For example, unusual geological substrates and landforms, including mountains, are well known for harboring species with restricted geographical ranges (Van der Werff 1992; Tuomisto and Poulsen 1996; Printaud and Jaffré 2001; de Klerk et al. 2002). Likewise, landscapes covering extensive areas, whether as a result of natural or anthropogenic processes, harbor species that have large geographical ranges (Terborgh and Winter 1983; Duncan, Blackburn, and Veltman 1999). And species' distributions are known to cluster in space, allowing the identification of geographical regions with unique characteristics (Hagmeier and Sults 1964).

Here we focus on North American birds and butterflies to address three questions. Does the distribution of geographical range sizes for these two taxa exhibit multiple modes? Do the distributions of geographical range sizes of these two unrelated volant taxa exhibit similarities? And are the modes and discontinuities in the size distribution of geographical ranges related in a meaningful way with landscape attributes?

METHODS

We restricted our analysis to North America defined as the continental mass that includes the United States, Canada, and Greenland; in some few instances, we included species whose geographical range extends into northern Mexico (fig. 7.1). The area occupied by the first three countries totals approximately 21.5 by 10^6 km^2, representing 14% of the Earth's land surface. Within this area, we included only those species whose geographical range falls completely within the boundaries described. This means that year-round residents and intracontinental migrants, but not intercontinental migrants, were included in our study. Although such restriction substantially decreased our sample size, it generated a more homogeneous group of species whose geographical ranges were more likely to be influenced by processes shaping North America as described earlier. It is well known that the majority of intercontinental migrants reported in North America belong to taxa that originated in the Neotropics (Levey and Stiles 1992). In addition, their geographical ranges are depicted as highly disjunct; breeding and wintering grounds do not overlap, raising the issue of how to measure their range size. In total, 136 species of birds and 288 species of butterflies met the criteria described here (see appendixes 7.1 and 7.2).

FIGURE 7.1 Map of North America showing main physiographic units (black lines) and Bailey's ecoregions (white lines). Numbers correspond to physiographic units: Pacific Mountain System (1), Intermontane Plateaus (2), Rocky Mountain System (3), Interior Plains (4), Interior Lowlands (5), Interior Highlands (6), Appalachian Highlands (7), Piedmont (8), Atlantic Plain (9), and Canadian Shield (10).

DISCONTINUITY IN GEOGRAPHICAL RANGE SIZE

We used published range maps of North American birds (National Geographic Society 2002) and butterflies (Scott 1986) to obtain geographical range-size data. Therefore, we express geographical range size in terms of extent, or the total area over which a species has been recorded, irrespective of range structure and shape (Brown, Stevens, and Kaufman 1996; Gaston and Blackburn 2000). For intracontinental migrants, we summed the breeding and wintering ranges to obtain a single figure for the size of their geographical range. We followed a three-step procedure to estimate the size of the geographical ranges. First, we scanned (600 dpi) the published range maps and processed the digital maps to eliminate pixels representing political boundaries and labels. Second, we used a clustering algorithm based on an eight-neighborhood rule to identify and measure the number of pixels in each cluster of the black-and-white range map images (Imagine, ERDAS). Last, we converted the number of pixels into metric units based on a model that predicts area (km^2) from pixel number. For this purpose, we selected from the range maps those features with known areas, such as states and provinces of the United States and Canada, and processed them as described here. This selection was necessary because the scale of the range maps differed between butterflies and birds, as well as within birds.

Geographical range-size data obtained in this fashion may have some limitations that need to be addressed. First, range maps can be generated using different methods that reflect predicted distribution based on habitat preferences, or the actual distribution based on field or museum observations or both (Brown, Stevens, and Kaufman 1996). Whereas we know that this latter method was used to generate the butterfly map ranges (Scott 1986), we do not know how the bird maps were prepared. Second, these maps, in contrast to those derived from coordinated large-scale censuses, are likely to provide a relatively crude estimate of the real size of geographical ranges and do not reflect the structure of the geographical range (Maurer 1994). However, maps derived from coordinated efforts are available only for limited taxa or geographical regions or both. Third, range maps generated by different authors are likely to be based on maps differing in terms of their projection or scale or both, as was the case with the bird and butterfly maps we used. Such difference may introduce an important source of error when data sets based on different maps are compared. Last, the size of small geographical ranges may be underestimated because of the small scale of the base maps. In spite of these limitations, range maps represent the best source of information available to estimate sizes of geographical ranges and make comparisons across taxa.

We used the Gap Rarity Index (GRI) to identify aggregations (or modes) and discontinuities (or gaps) in the size distribution of geographical ranges (Restrepo, Renjifo, and Marples 1997). The GRI method tests whether discontinuities in an observed distribution of rank-ordered data are unlikely in data sampled from a continuous unimodal log-normal distribution fit to the observed data. First, a continuous unimodal distribution is obtained by constructing a normal kernel density estimate that uses the smallest window width (h) that smoothes the observed frequency distribution (Silverman 1986). Second, absolute gaps in a variable of interest are measured, $d_I = s_{i+1} - s_i$, where s_i is the $\log_{10}$ of ith geographical range size in rank-size-ordered data, and their significance was tested based on the index, D_i. This index is a statistic measuring the proportion of simulated absolute gaps smaller than the observed, and it is obtained by sampling the continuous unimodal distribution ten thousand times (Restrepo, Renjifo, and Marples 1997).

CLUSTERS OF SPECIES IN GEOGRAPHICAL SPACE

The geographical range of a species reflects the gamut of abiotic and biotic conditions influencing a species' life span in space and time and therefore is indicative of a species' history. Broad-scale subdivisions of land based on terrain structure and geology—that is, physiographic units—are ideal to classify species in geographical space. Such a map exists for the United States (Fenneman and Johnson 1946; Vigil, Pike, and Howell 2002), but not for the entire region considered in this study (but see Barton, Howell, and Vigil 2003). Instead, we used Bailey's (1998) map of North American ecosystems, or ecoregions, which reflects physiographic units to some extent. In addition, the resolution of the ecoregions was fine enough to allow a detailed characterization of birds and butterflies' geographical ranges, especially those restricted to small areas. We refer to major physiographic provinces (Fenneman and Johnson 1946) to report our results (fig. 7.1).

After differentiating ecoregions represented on both coasts into east and west to account for the different origins of the land (King and Beikman 1974), we identified for each species the ecoregions falling within the limits of the range maps. We recorded ecoregions as present or absent irrespective of whether the ecoregion was widely or narrowly represented. We ran a cluster analysis using Sorensen's index as a distance measure to avoid the double zero problem and using flexible beta ($\beta = -0.25$) as the group linkage method to build dendrograms (Legendre and Legendre 2003). We used branches in the dendrogram at 50% of similarity to distinguish clusters of species occupying similar geographical regions, hereafter referred to as *zooregions* (Hagmeier and Sults 1964). The misplacement of some species in the clusters was unavoidable, in part because of the scale of the range maps and our inability to identify ecoregions within the range maps. This misplacement may have resulted in the inclusion of some ecoregions that were not really represented within the boundaries of the published maps. We classified each species according to aggregation in geographical range size and zooregion, and we used chi-square tests to evaluate whether affiliation to a given aggregation was independent from affiliation to a given zooregion.

RESULTS

The geographical ranges of birds and butterflies exhibited a right-skewed distribution and overlapped over a wide range of values (4.9 by 10^4 to 1.3 by 10^7 km^2), yet butterflies exhibited the smallest and birds the largest geographical ranges (3.8 by 10^4 km^2 and 1.51 by 10^7 km^2, respectively) (fig. 7.2). The observed distribution of bird and butterfly geographical range sizes did not differ

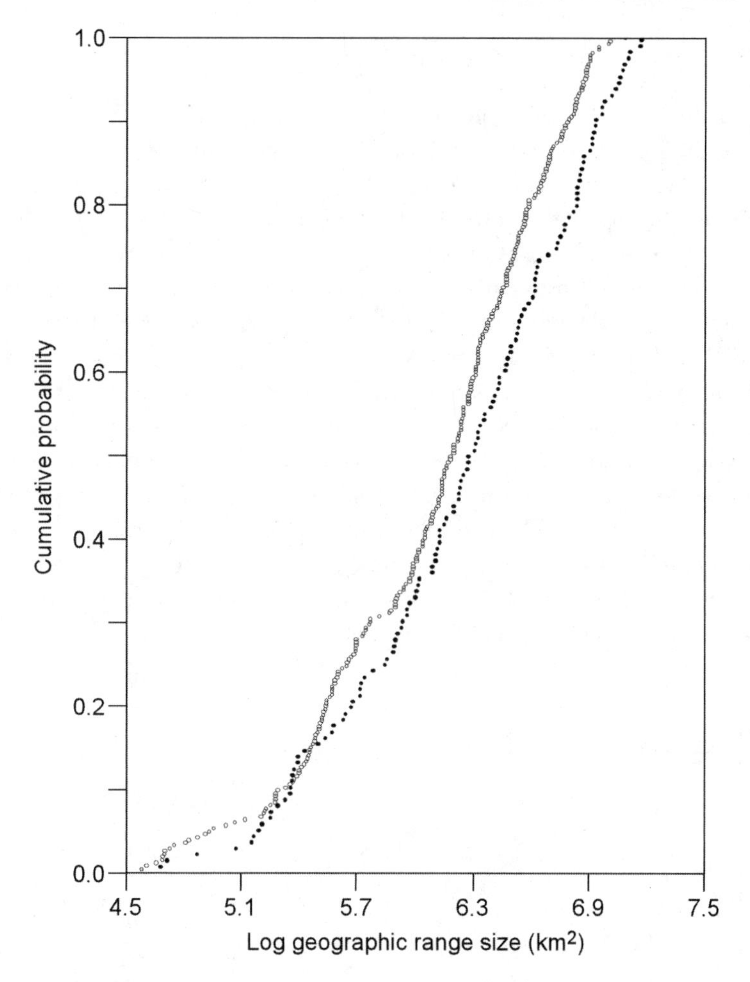

FIGURE 7.2 Observed cumulative distribution function of geographical range size of North American birds (●) and butterflies (O).

($K-S=0.1327$, $P=0.07$) (fig. 7.2). When compared against an expected log-normal distribution, however, the observed distribution of butterfly geographical ranges ($K-S=0.08$, $P=0.03$), but not of bird geographical ranges ($K-S=0.06$, $P=0.69$), was significantly different (fig. 7.3). This difference suggests that properties other than the mean and standard deviation, the two parameters that describe the shape of log-normal distributions, may be responsible for the similarities between the two observed distributions. One possibility is that the observed distributions exhibit patterns of discontinuity and aggregation.

DISCONTINUITY IN GEOGRAPHICAL RANGE SIZE

An examination of the cumulative density curves shows a large discontinuity at approximately 5.8 km^2 (log-transformed value or 6.0-by-10^5-km^2 untransformed value) in both data sets, hereafter referred to as the 6GAP (figs. 7.3 and 7.4). Furthermore, changes in the slope of the cumulative density curves of the observed data in several regions suggested additional discontinuity when compared to the expected log-normal cumulative density curves. We identified seven aggregations in the distribution of geographical range sizes of butterflies ($P=0.006$) and birds ($P=0.002$) using the GRI method (fig. 7.4). This analysis confirmed the presence of the discontinuity found at approximately 5.8 km^2 (log-transformed value) as well as of other discontinuities already observed in the cumulative density curves of the two taxa. Below the 6GAP, we identified three and two aggregations in the bird and butterfly data sets, respectively. In this region, there seems to be a match in aggregation 1 in each of the data sets and between aggregation 2 for butterflies and aggregations 2 and 3 for birds. Above the 6GAP, we found four and five aggregations in the bird and butterfly data sets, respectively. Aggregations 4 and 5 for birds seem to match aggregation 3 for butterflies, but any resemblance thereafter is less obvious.

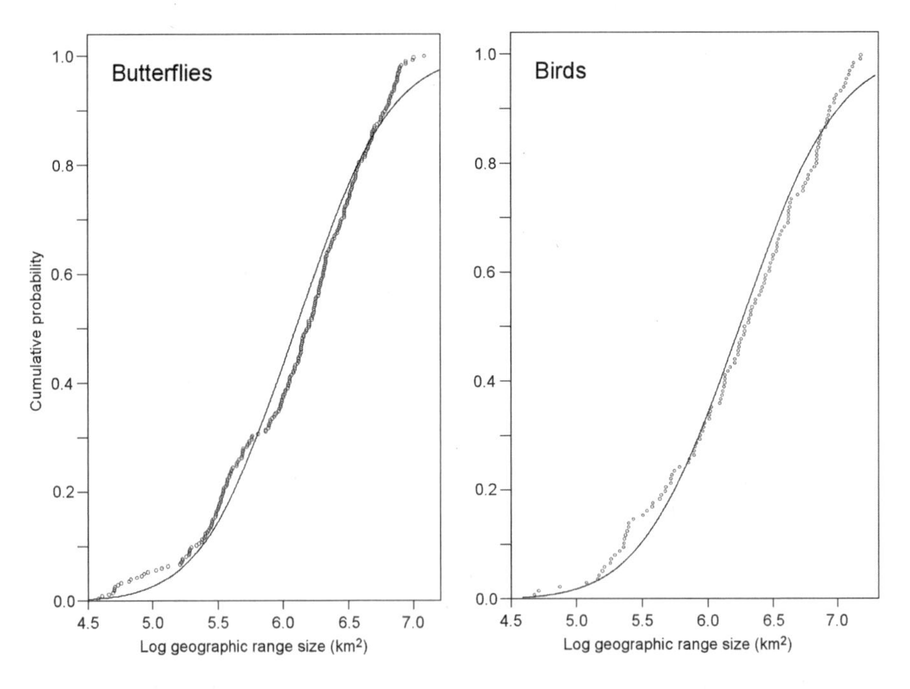

FIGURE 7.3 Observed (O) and expected (black line) cumulative distribution functions of geographical range size of North American butterflies and birds. The expected cumulative distribution functions correspond to a log-normal distributions with parameters estimated from the data. There is a large discontinuity in the observed distributions at approximately 5.8 km2 (log-transformed data). We refer to this discontinuity as the 6GAP.

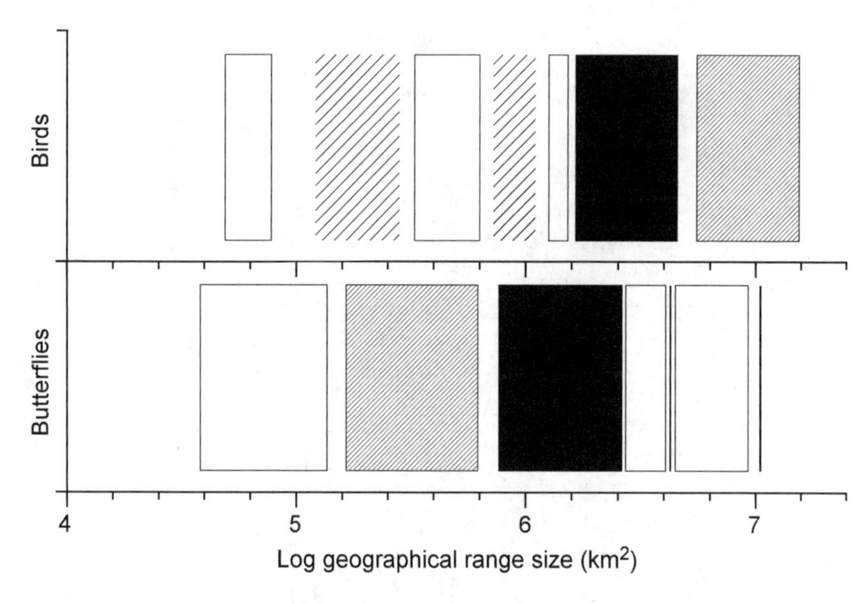

FIGURE 7.4 Identified discontinuities and aggregations in the distribution of geographical range size of North American birds and butterflies resulting from the GRI analysis. Changes in color and hatching from white to black indicate a greater percentage of species falling within each aggregation, with black indicating the largest percentage.

CLUSTERS OF SPECIES IN GEOGRAPHICAL SPACE

For birds, we identified six major clusters, or zooregions (fig. 7.5). Zooregion 1 included species restricted to small areas in the Pacific Mountain System (California and Baja California) and small areas of the Intermontane Plateaus (southwestern United States). Zooregion 2 was represented by species found in the Interior Plains (southwestern United States), in some instances reaching into the southern Rocky Mountain System and the Intermontane Plateaus. Species found in the Atlantic Plain and increasingly extending into the Piedmont, Appalachian Highlands, and Interior Highlands were grouped in zooregion 3. A small group of species found in the Pacific Mountain System of the northwestern United States (including Alaska) and western Canada defined zooregion 4. Species found mostly in the Rocky Mountain System and the Intermontane Plateaus defined zooregion 5. Finally, species found in the Pacific Mountain System, Intermontane Plateaus, and Rocky Mountain System from Alaska to northern California, Arizona, and New Mexico and then extending into the Canadian Shield (Laurentian Upland and Lowland), Interior Highlands, Appalachian Highlands, and Atlantic Plain were grouped in zooregion 6.

Butterflies clustered in geographical space in a similar, but not identical, manner as birds clustered (fig. 7.6). We identified seven major zooregions, the first

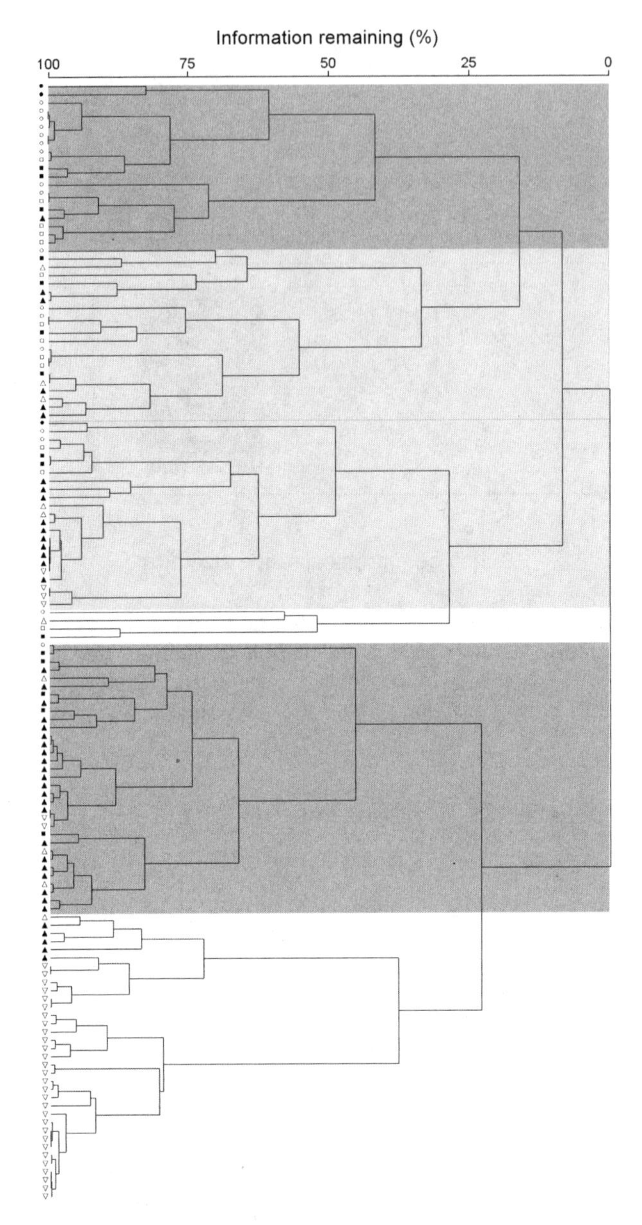

FIGURE 7.5 Dendrogram showing clusters of North American birds based on the occurrence of ecoregions within their geographical ranges. Each species is also identified with a symbol indicating aggregation number affiliation: aggregation 1 (●), aggregation 2 (○), aggregation 3 (□), aggregation 4 (■), aggregation 5 (△), aggregation 6 (▲), and aggregation 7 (▽). Species' clusters are identified with the large gray and white boxes.

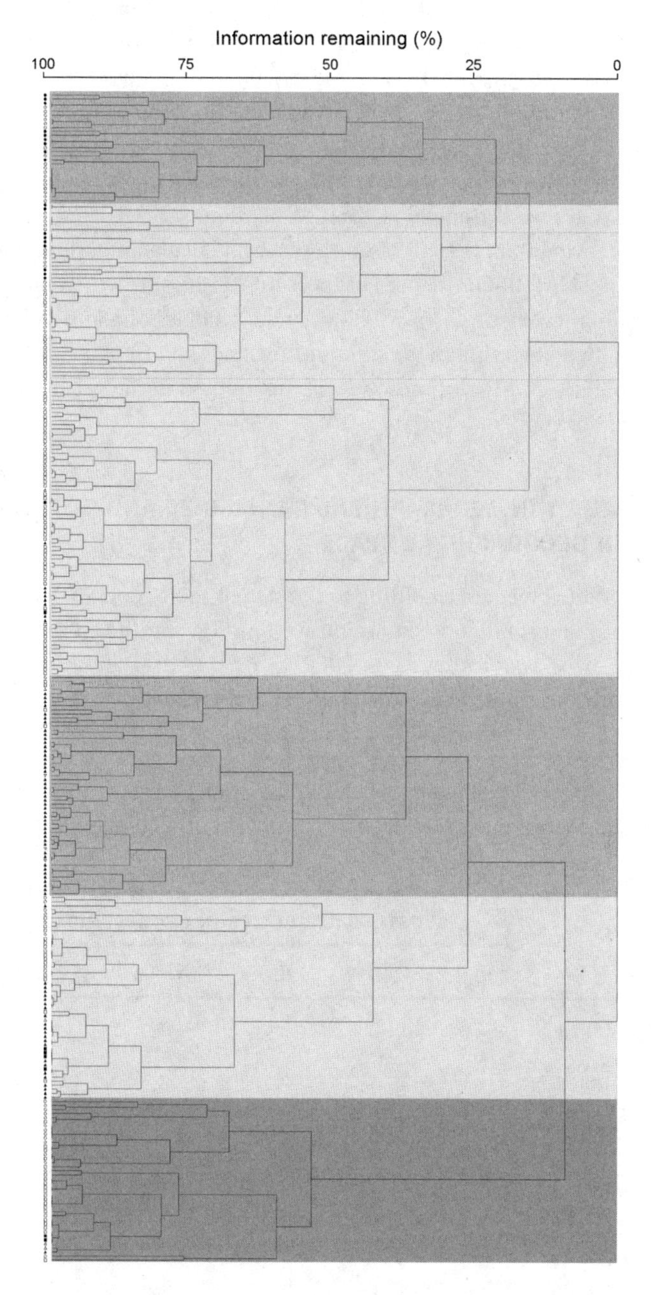

FIGURE 7.6 Dendrogram showing clusters of North American butterflies based on the occurrence of ecoregions within their geographical ranges. Each species is also identified with a symbol indicating aggregation number affiliation: aggregation 1 (●), aggregation 2 (○), aggregation 3 (□), aggregation 4 (■), aggregation 5 (△), aggregation 6 (▲), and aggregation 7 (▽). Species' clusters are identified with the large gray and white boxes.

composed of species restricted to sites in the Pacific Mountain System (California). Zooregion 2 grouped butterfly species found mostly in the Intermontane Plateaus (southwestern United States and extending into northern Mexico). Species found in the Intermontane Plateaus and Interior Plains of the southwestern United States were grouped in zooregion 3. Species found in the Rocky Mountain System, northern Intermontane Plateaus, and Pacific Mountain System of the United States were grouped in zooregion 4. Species restricted to small areas in the Pacific Mountain System of northwestern Canada and extending into the northern Pacific Mountain System of the United States, Intermontane Plateaus, Rocky Mountain System, Interior Plains, and Canadian Shield were grouped in zooregion 5. Species found in the Interior Lowlands and Appalachian Highlands were in zooregion 6. Finally, those species distributed in the Atlantic Plain were grouped in zooregion 7.

DISCONTINUITY IN GEOGRAPHICAL RANGE SIZE AND CLUSTERS OF SPECIES IN GEOGRAPHICAL SPACE

We classified bird species according to the size of their geographical range (aggregation number) and zooregion, and we found a significant association between these two variables ($\chi^2 = 131.9$, $df = 20$, $P = 0.0001$) (table 7.1; fig. 7.5). An examination of the post hoc individual cell values revealed that species with

TABLE 7.1 Number of North American Bird Species Classified in Terms of Their Geographical Range Size (Aggregation Number) and Zooregion (Cluster of Species in Geographical Space)

		CLUSTERS OF SPECIES IN GEOGRAPHICAL SPACE					
		ZOO_1	ZOO_2	ZOO_4	ZOO_5	ZOO_6	ZOO_3
GEOGRAPHICAL RANGE SIZEE	AGGREGATION 1	2	0	0	0	0	1
	AGGREGATION 2	9	4	1	1	0	2
	AGGREGATION 3	5	5	1	0	0	2
	AGGREGATION 4	3	4	1	5	0	2
	AGGREGATION 5	0	3	1	3	1	2
	AGGREGATION 6	1	5	0	22	5	10
	AGGREGATION 7	0	0	0	2	29	4

Note: Zooregions 1 and 2 and aggregations 1 and 2 were pooled to carry out the statistical analysis described in the text.

TABLE 7.2 Number of North American Butterflies Classified in Terms of Their Geographical Range Size (Aggregation Number) and Zooregion (Cluster of Species in Geographical Space)

	CLUSTERS OF SPECIES IN GEOGRAPHICAL SPACE						
GEOGRAPHICAL RANGE SIZEE	ZOO_1	ZOO_2	ZOO_3	ZOO_4	ZOO_5	ZOO_6	ZOO_7
AGGREGATION 1	9	2	7	0	0	0	0
AGGREGATION 2	18	5	19	7	1	6	14
AGGREGATION 3	0	0	10	55	4	16	20
AGGREGATION 4	0	0	0	2	0	4	2
AGGREGATION 5	0	0	0	1	1	1	2
AGGREGATION 6	0	0	0	8	45	23	1
AGGREGATION 7	0	0	0	0	4	0	0

Note: Zooregions 1 and 2 and aggregations 6 and 7 were pooled to carry out the statistical analysis described in the text.

small geographical ranges (aggregations 1 and 2) were represented more often than expected in zooregions 1 and 2. Likewise, species in aggregations 6 and 7 were found more often than expected in zooregions 5 and 6, respectively.

The butterfly zooregions were significantly associated with geographical range size ($\chi^2 = 305.6$, $df = 25$, $P = 0.0001$) (table 7.2; fig. 7.6). We found that species with small geographical ranges (aggregations 1 and 2) were found more often than expected in zooregions 1 through 3. These ranges were followed by medium-size geographical ranges (aggregation 3) found more often than expected in zooregion 4. The largest geographical ranges (aggregations 6 and 7) were found more often than expected in zooregions 5 and 6.

DISCUSSION

The distribution of geographical range sizes for North American birds and butterflies was discontinuous, characterized by aggregations and discontinuity (gaps). Moreover, the location of aggregations and gaps in the two data sets exhibited important similarities. Further, there is a strong association between aggregations in the size distribution of geographical ranges and clusters of species in geographical space. Taken altogether, these three findings support the idea that discontinuities and aggregations in the distribution of geographical range size may reflect large-scale spatial attributes as predicted by the TDH (Holling 1992).

Although this is the first time to our knowledge that Holling's TDH has been tested in a biogeographical context, some earlier work has shown that when geographical range size is expressed in terms of site occupancy, bimodal distributions arise (Hanski 1982). Specifically, there is a high frequency of species that are locally sparse and regionally uncommon and of species that are locally abundant and regionally common; in between these two extremes fall species with intermediate site occupancies. It has been proposed that such patterns may result from biological processes (low environmental heterogeneity) or artefactual effects (small sampling extents) (for a recent review, see McGeoch and Gaston 2002). Yet the distribution of geographical range size of mammals in North America (Simpson 1964) and of birds in the Neotropics (Gaston and Blackburn 1997), two large and heterogeneous regions, seems to be bimodal.

The 6GAP at approximately 600,000 km^2 (untransformed value) was a distinctive feature in both the bird and the butterfly data sets. In the bird data, we clearly identified three aggregations ($n = 33$ species, or 24% of the total) below this value, and most of the member species were found in the Pacific Mountain System (California) and the southern Intermontane Plateaus (Texas, New Mexico, Arizona, and northern Mexico). A very small fraction of the species falling within these three aggregations were restricted to the Atlantic Plains, either to the central sand dunes of Florida or to a narrow strip along the coastline, extending increasingly inwards. In the butterfly data, we identified two aggregations below the 6GAP ($n = 88$ species, or 31% of the total), and, as for birds, these aggregations included mostly species found in the southern Pacific Mountain System (California), the Intermontane Plateaus, and the Interior Plains. Unlike for birds, however, we found many more butterfly species with small ranges centered in the Interior Plains (Texas) and the Atlantic Plains. In fact, this finding may have contributed to our recognition of an additional zooregion for butterflies, zooregion 7. For the most part, however, the 6GAP separated butterfly species and bird species with small geographical ranges (less than 1% of the size of the North American continent) that seem to be associated with complex or relatively recent landforms, or both (Fenneman and Johnson 1946; King and Beikman 1974). These landforms are ecotonal in character in two ways. First, they developed along the margins of the stable core of North America and are relatively new from a geological perspective. Second, they are currently influenced by a subtropical climate.

In the bird data set, we identified four aggregations above the 6GAP. Aggregations 4 and 5 had few species distributed more or less evenly among the six zooregions, whereas aggregations 6 and 7 included the largest number of species (not only above the gap, but overall) and were characteristically associated with zooregions 5 and 6, respectively. The attributes of the geographical ranges of species belonging to aggregations 4 and 5 are thus apparently "transitional" in character in that these species, unlike species with smaller or larger ranges, are not strongly associated with any zooregion. This conclusion is illustrated by the following two examples. First, although most birds considered in our analyses have their geographical ranges within the North American continent as defined in this work, we included a few that extended south to central Mexico. These species were found in aggregations 4 and 5, and the inclusion of this additional piece of land apparently introduced a source of heterogeneity not found within the area that we defined as the North American continent. In other words, this landform left an imprint on the size distribution of geographical ranges. Second, within a given zooregion some species had "unusual" geographical ranges. For example, three species in aggregation 4 found in the Pacific Mountain System had narrow but very long geographical ranges extending along most of the coast of Canada and the United States *(Sphyrapicus ruber* and *Calypte anna)* or along the coast of the United States and Mexico *(Calypte costae)*. Other species were all-year residents *(Strix occidentalis* and *Lagopus leucurus)* that have disjunct populations such that their geographical ranges include dissimilar ecoregions. In contrast to aggregations 4 and 5, aggregation 6 had the largest number of species (32%), the vast majority of which had geographical ranges within the Rocky Mountain System and the Intermontane Plateaus of the United States. Aggregation 7 had the second-largest number of species (26%), but these species, unlike most of the species in aggregation 6, were found distributed across the continent in an east–west direction either centered in the Canadian Shield physiographic unit or entering the Rocky Mountain System and the Appalachian Highlands. Only two species had ranges spanning most of the study area *(Junco hyemalis* and *Colaptes auratus)*, but these ranges were nevertheless smaller than the total area.

For butterflies, we found a large aggregation above the 6GAP (106 species, or 37% of the total) that included species with geographical ranges centered in the Rocky Mountain System and Intermontane Plateaus. This aggregation mirrors aggregation 6 for birds. Aggregations 4 and 5 had very few butterfly species and, as found for birds, had "unusual" geographical ranges for the zooregion in which they fell. They marked a transition between aggregations 3 and 6, the latter including species with geographical ranges running predominantly in an east–west direction, as was the case for birds.

Two non–mutually exclusive explanations may account for these results. First, aggregations and discontinuities in the distribution of geographical range size are the result of changes in the shape of the geographical ranges, which in turn are influenced, if not constrained, by the structure of the landforms where the species originated. Second, range expansion or contraction over a species' life span may have translated into changes of geographical range shape and size, and therefore into changes in the overall distribution of range sizes at a continental scale. We speculate that underlying these two explanations is a problem of geographical range allometry and, more specifically, of the occurrence of scale breaks most likely resulting from the ways in which landscapes are structured at large scales. These hypotheses can be easily tested by examining the distribution of geographical range size of plant and animal assemblages from other continents of equivalent area and range of climate, but of different structure.

The qualitative similarities between birds and butterflies above the 6GAP, however, were not matched in terms of the correspondence of aggregations and discontinuities. One reason for this result may be that differences between the bird and butterfly maps were ultimately reflected in the size and shape of the ranges. Also, the level of detail in the two sets of maps may have affected our ability to discern which ecoregions were found within the ranges. Alternatively, the lack of correspondence of aggregations and discontinuities above the 6GAP in both data sets may reflect real differences in the way birds and butterflies, two taxa that differ greatly in size, perceive and exploit resources. The occurrence of a larger number of butterflies in each aggregation-zooregion combination may provide support to this idea.

The simultaneous examination of geographical range size and ecoregions provides tremendous insight into the processes underlying the distribution of attributes used to characterize species assemblages. In particular, the distribution of geographical range size has previously been characterized in most instances as a continuous, unimodal, right-skewed distribution that may become normal or slightly left skewed when log transformed. Instead, we found several aggregations of varying size clearly associated with landscape attributes. These findings may have implications in terms of how we define endemic species, how we predict which geographical range sizes are likely to expand or contract, and perhaps which areas deserve special conservation status because many species seem to originate in them.

APPENDIX 7.1 North American Bird Species Included in the Study of Discontinuities in Geographical Range Size

FAMILY	SPECIES	GEOGRAPHICAL RANGE SIZE (KM^2)	AGGREGATION NUMBER
CATHARTIDAE	*Gymnogyps californianus*	48,893.40	1
CORVIDAE	*Aphelocoma coerulescens*	52,664.61	1
CORVIDAE	*Pica nuttalli*	76,695.89	1
PHASIANIDAE	*Tympanuchus pallidicinctus*	121,726.58	2
EMBERIZIDAE	*Ammodramus caudacutus*	146,754.99	2
EMBERIZIDAE	*Plectrophenax hyperboreus*	150,598.00	2
EMBERIZIDAE	*Aimophila carpalis*	160,181.28	2
FRINGILLIDAE	*Carduelis laurencei*	165,634.29	2
FRINGILLIDAE	*Leucosticte australis*	183,919.68	2
EMBERIZIDAE	*Pipilo aberti*	185,375.69	2
PARIDAE	*Poecile sclateri*	199,200.96	2
EMBERIZIDAE	*Ammodramus maritimus*	216,923.62	2
MIMIDAE	*Toxostoma redivivum*	231,778.70	2
PARIDAE	*Baeolophus inornatus*	234,654.83	2
MIMIDAE	*Toxostoma lecontei*	236,188.01	2
EMBERIZIDAE	*Agelaius tricolor*	238,812.84	2
PICIDAE	*Picoides nuttallii*	245,561.42	2
POLIOPTILIDAE	*Polioptila californica*	254,167.75	2
TIMALIIDAE	*Chamaea fasciata*	254,762.01	2
PICIDAE	*Picoides albolarvatus*	276,036.31	2
EMBERIZIDAE	*Pipilo crissalis*	323,744.90	3
EMBERIZIDAE	*Quiscalus major*	349,633.87	3
PICIDAE	*Colaptes chrysoides*	383,256.72	3

APPENDIX 7.1 *Continued*

FAMILY	SPECIES	GEOGRAPHICAL RANGE SIZE (KM^2)	AGGREGATION NUMBER
PHASIANIDAE	*Tympanuchus cupido*	387,283.86	3
PARIDAE	*Baeolophus wollweberi*	437,230.08	3
ODONTOPHORIDAE	*Oreortyx pictus*	447,287.38	3
CORVIDAE	*Corvus caurinus*	480,633.52	3
MIMIDAE	*Toxostoma bendirei*	486,047.02	3
CORVIDAE	*Aphelocoma ultramarina*	530,984.16	3
PICIDAE	*Melanerpes uropygialis*	531,370.05	3
EMBERIZIDAE	*Aimophila aestivalis*	537,033.51	3
ODONTOPHORIDAE	*Callipepla gambelii*	563,372.01	3
STRIGIDAE	*Micrathene whitneyi*	617,440.32	3
VIREONIDAE	*Vireo vicinitor*	721,610.06	4
FRINGILLIDAE	*Leucosticte atrata*	734,140.31	4
SITTIDAE	*Sitta pusilla*	791,941.02	4
STRIGIDAE	*Strix occidentalis*	796,937.33	4
TROCHILIDAE	*Calypte costae*	810,120.02	4
PICIDAE	*Picoides borealis*	828,931.32	4
PARIDAE	*Baeolophus rufescens*	879,115.83	4
TROCHILIDAE	*Calypte anna*	887,002.66	4
PHASIANIDAE	*Centrocercus urophasianus*	932,101.98	4
PARIDAE	*Baeolophus griseus*	936,687.33	4
ODONTOPHORIDAE	*Callipepla californica*	959,438.99	4
EMBERIZIDAE	*Ammodramus bairdii*	1,037,495.49	4
POLIOPTILIDAE	*Polioptila melanura*	1,040,316.48	4
PICIDAE	*Sphyrapicus ruber*	1,060,857.59	4

APPENDIX 7.1 *Continued*

FAMILY	SPECIES	GEOGRAPHICAL RANGE SIZE (KM^2)	AGGREGATION NUMBER
MIMIDAE	*Toxostoma crissale*	1,079,588.16	4
EMBERIZIDAE	*Calcarius mccownii*	1,252,168.22	5
CORVIDAE	*Gymnorhinus cyanocephalus*	1,266,450.19	5
EMBERIZIDAE	*Ammodramus henslowii*	1,303,476.69	5
CORVIDAE	*Corvus ossifragus*	1,309,994.28	5
PHASIANIDAE	*Lagopus leucurus*	1,335,954.86	5
EMBERIZIDAE	*Ammodramus nelsoni*	1,358,267.16	5
EMBERIZIDAE	*Pipilo fuscus*	1,369,034.07	5
SITTIDAE	*Sitta pygmaea*	1,370,127.16	5
EMBERIZIDAE	*Aimophila ruficeps*	1,440,839.10	5
CORVIDAE	*Corvus cryptoleucus*	1,502,145.14	5
MIMIDAE	*Toxostoma curvirostre*	1,624,583.52	6
REMIZIDAE	*Auriparus flaviceps*	1,628,039.49	6
PICIDAE	*Sphyrapicus thyroideus*	1,720,874.50	6
MOTACILLIDAE	*Anthus spraguelii*	1,729,808.87	6
TROGLODYTIDAE	*Campylorhynchus brunneicapillus*	1,748,492.01	6
EMBERIZIDAE	*Aimophila cassinii*	1,767,703.44	6
CORVIDAE	*Aphelocoma californica*	1,811,037.86	6
EMBERIZIDAE	*Calcarius pictus*	1,900,925.60	6
EMBERIZIDAE	*Amphispiza belli*	1,932,016.92	6
CORVIDAE	*Nucifraga columbiana*	1,933,806.13	6
FRINGILLIDAE	*Carpodacus cassinii*	2,067,875.81	6
PICIDAE	*Melanerpes lewis*	2,086,872.01	6

APPENDIX 7.1 *Continued*

FAMILY	SPECIES	GEOGRAPHICAL RANGE SIZE (KM²)	AGGREGATION NUMBER
EMBERIZIDAE	*Zonotrichia querula*	2,149,974.72	6
EMBERIZIDAE	*Calcarius ornatus*	2,163,255.55	6
PARIDAE	*Poecile carolinensis*	2,199,436.03	6
PHASIANIDAE	*Dendragapus obscurus*	2,330,278.15	6
PARIDAE	*Poecile gambeli*	2,343,351.63	6
TURDIDAE	*Sialia mexicana*	2,512,742.54	6
TROCHILIDAE	*Archilochus alexandri*	2,570,849.52	6
EMBERIZIDAE	*Zonotrichia atricapilla*	2,637,925.13	6
EMBERIZIDAE	*Dendroica pinus*	2,735,331.67	6
MIMIDAE	*Oreoscoptes montanus*	2,775,922.45	6
EMBERIZIDAE	*Calamospiza melanocorys*	2,807,039.10	6
EMBERIZIDAE	*Pipilo chlorurus*	2,985,712.21	6
PICIDAE	*Picoides scalaris*	3,008,385.58	6
TROGLODYTIDAE	*Catherpes mexicanus*	3,059,461.11	6
TURDIDAE	*Ixoreus naevius*	3,182,726.29	6
STRIGIDAE	*Otus kennicotis*	3,196,894.34	6
EMBERIZIDAE	*Spizella breweri*	3,399,900.45	6
EMBERIZIDAE	*Ammodramus leconteii*	3,439,489.14	6
PICIDAE	*Melanerpes carolinus*	3,451,570.76	6
PICIDAE	*Sphyrapicus nuchalis*	3,523,635.80	6
PARIDAE	*Baeolophus bicolor*	3,621,328.15	6
FRINGILLIDAE	*Leucosticte tephrocotis*	3,748,610.12	6
ACCIPITRIDAE	*Buteo regalis*	3,944,000.16	6
TROGLODYTIDAE	*Thryothorus bewickii*	4,173,406.16	6

APPENDIX 7.1 *Continued*

FAMILY	SPECIES	GEOGRAPHICAL RANGE SIZE (KM^2)	AGGREGATION NUMBER
CAPRIMULGIDAE	*Phalaenoptilus nuttallii*	4,224,045.37	6
PHASIANIDAE	*Tympanuchus phasianellus*	4,228,093.52	6
FALCONIDAE	*Falco mexicanus*	4,231,703.52	6
EMBERIZIDAE	*Spizella pusilla*	4,306,910.57	6
TURDIDAE	*Sialia cucorroides*	4,335,942.32	6
PICIDAE	*Melanerpes erythrocephalus*	4,454,873.42	6
STRIGIDAE	*Otus asio*	4,980,413.52	6
MIMIDAE	*Toxostoma rufum*	5,475,720.86	7
PICIDAE	*Picoides arcticus*	5,535,010.53	7
PHASIANIDAE	*Meleagris gallipavo*	5,745,221.01	7
TURDIDAE	*Myadestes townsendi*	5,981,241.13	7
EMBERIZIDAE	*Pipilo maculatus*	6,054,404.84	7
STRIGIDAE	*Strix varia*	6,345,958.06	7
PARIDAE	*Poecile hudsonicus*	6,680,823.14	7
PICIDAE	*Dryocopus pileatus*	6,966,758.31	7
FRINGILLIDAE	*Coccothraustes vespertinus*	6,999,336.80	7
CORVIDAE	*Cyanocitta cristata*	7,008,286.42	7
LANIIDAE	*Lanius ludovocianus*	7,023,979.92	7
PHASIANIDAE	*Bonasa umbellus*	7,180,719.96	7
SITTIDAE	*Sitta carolinensis*	7,250,946.56	7
PHASIANIDAE	*Dendragapus canadensis*	7,378,870.52	7
STRIGIDAE	*Aegolius acadicus*	7,496,170.88	7
FRINGILLIDAE	*Carpodacus purpureus*	7,563,130.25	7
CORVIDAE	*Perisoreus canadensis*	8,139,837.84	7

APPENDIX 7.1 *Continued*

FAMILY	SPECIES	GEOGRAPHICAL RANGE SIZE (KM^2)	AGGREGATION NUMBER
PARIDAE	*Poecile atricapillus*	8,321,408.04	7
EMBERIZIDAE	*Quiscalus quiscala*	8,362,835.65	7
TROGLODYTIDAE	*Cistothorus palustris*	8,577,547.55	7
EMBERIZIDAE	*Euphagus carolinus*	8,711,199.08	7
EMBERIZIDAE	*Zonotrichia albicollis*	8,780,431.01	7
EMBERIZIDAE	*Passerella iliaca*	9,419,367.36	7
EMBERIZIDAE	*Spizella arborea*	9,498,511.88	7
FRINGILLIDAE	*Carduelis tristis*	9,808,459.52	7
REGULIDAE	*Regulus satrapa*	10,668,588.22	7
CORVIDAE	*Corvus brachyrhynchus*	11,210,082.59	7
PICIDAE	*Picoides pubescens*	11,531,589.52	7
SITTIDAE	*Sitta canadensis*	11,722,684.16	7
EMBERIZIDAE	*Melospiza melodia*	12,248,430.24	7
EMBERIZIDAE	*Zonotrichia leucophrys*	12,390,931.26	7
FRINGILLIDAE	*Carduelis pinus*	13,028,008.42	7
ACCIPITRIDAE	*Haliaethus leucocephalus*	13,298,938.88	7
PICIDAE	*Colaptes auratus*	14,994,305.15	7
EMBERIZIDAE	*Junco hyemalis*	15,068,746.85	7

APPENDIX 7.2 North American Butterfly Species Included in the Study of Discontinuities in Geographical Range Size

FAMILY	SPECIES	GEOGRAPHICAL RANGE SIZE (KM^2)	AGGREGATION NUMBER
PIERIDAE	*Colias behrii*	38,451.08	1
HESPERIIDAE	*Agathymus evansi*	41,385.10	1
HESPERIIDAE	*Polites mardon*	46,403.81	1
LYCAENIDAE	*Lycaena hermes*	49,569.46	1
LYCAENIDAE	*Plebejus emigdionis*	50,495.99	1
HESPERIIDAE	*Hesperia miriamae*	51,036.47	1
LYCAENIDAE	*Callophrys dumetorum*	51,499.74	1
NYMPHALIDAE	*Chlosyne chinatiensis*	54,588.18	1
LYCAENIDAE	*Plebejus neurona*	57,290.56	1
HESPERIIDAE	*Amblyscirtes unnamed*	66,478.67	1
NYMPHALIDAE	*Speyeria adiaste*	68,872.21	1
HESPERIIDAE	*Atrytonopsis cestus*	77,133.79	1
HESPERIIDAE	*Thorybes diversus*	84,005.57	1
LYCAENIDAE	*Fixsenia polingi*	87,943.33	1
HESPERIIDAE	*Celotes limpia*	93,579.73	1
HESPERIIDAE	*Agathymus stephensi*	107,940.98	1
HESPERIIDAE	*Agathymus remingtoni*	120,140.32	1
HESPERIIDAE	*Agathymus alliae*	134,501.56	1
NYMPHALIDAE	*Phyciodes orseis*	165,154.33	2
HESPERIIDAE	*Agathymus polingi*	168,011.14	2
LYCAENIDAE	*Calephelis wrighti*	171,871.69	2
NYMPHALIDAE	*Coenonympha haydenii*	172,798.22	2
LYCAENIDAE	*Philotes sonorensis*	183,530.55	2

APPENDIX 7.2 *Continued*

FAMILY	SPECIES	GEOGRAPHICAL RANGE SIZE (KM²)	AGGREGATION NUMBER
NYMPHALIDAE	*Boloria natazhati*	193,336.34	2
HESPERIIDAE	*Agathymus neumoegeni*	195,266.62	2
HESPERIIDAE	*Atrytonopsis lunus*	195,652.67	2
HESPERIIDAE	*Atrytonopsis deva*	195,961.52	2
HESPERIIDAE	*Piruna polingii*	197,351.32	2
HESPERIIDAE	*Problema bulenta*	201,057.44	2
HESPERIIDAE	*Amblyscirtes fimbriata*	220,514.62	2
NYMPHALIDAE	*Speyeria diana*	231,710.21	2
HESPERIIDAE	*Amblyscirtes nereus*	240,203.42	2
HESPERIIDAE	*Euphyes arpa*	244,295.60	2
NYMPHALIDAE	*Chlosyne hoffmanni*	251,939.49	2
NYMPHALIDAE	*Boloria kriemhild*	259,351.75	2
NYMPHALIDAE	*Erebia vidleri*	259,660.59	2
HESPERIIDAE	*Euphyes berryi*	266,609.58	2
HESPERIIDAE	*Agathymus aryxna*	273,249.73	2
PIERIDAE	*Colias eurydice*	279,117.77	2
HESPERIIDAE	*Pyrgus xanthus*	285,912.33	2
LYCAENIDAE	*Satyrium auretorum*	286,761.65	2
LYCAENIDAE	*Philotiella speciosa*	292,398.06	2
LYCAENIDAE	*Callophrys lanoraieensis*	292,475.27	2
HESPERIIDAE	*Hesperia lindseyi*	296,953.51	2
HESPERIIDAE	*Atrytonopsis pittacus*	306,990.94	2
HESPERIIDAE	*Stinga morrisoni*	309,538.90	2
HESPERIIDAE	*Megathymus ursus*	312,472.92	2

APPENDIX 7.2 *Continued*

FAMILY	SPECIES	GEOGRAPHICAL RANGE SIZE (KM^2)	AGGREGATION NUMBER
LYCAENIDAE	*Euphilotes spaldingi*	314,557.61	2
HESPERIIDAE	*Hesperia columbia*	320,657.28	2
HESPERIIDAE	*Amblyscirtes phylace*	327,065.80	2
NYMPHALIDAE	*Chlosyne californica*	328,687.23	2
HESPERIIDAE	*Megathymus cofaqui*	333,242.68	2
HESPERIIDAE	*Nastra neamathla*	338,570.24	2
NYMPHALIDAE	*Euphydryas gillettii*	338,879.08	2
HESPERIIDAE	*Ochlodes agricola*	340,500.51	2
PAPILIONIDAE	*Papilio brevicauda*	346,831.81	2
NYMPHALIDAE	*Oeneis nevadensis*	352,854.27	2
HESPERIIDAE	*Hesperia dacotae*	356,405.98	2
PIERIDAE	*Anthocharis lanceolata*	357,255.30	2
HESPERIIDAE	*Agathymus mariae*	358,876.73	2
HESPERIIDAE	*Amblyscirtes cassus*	370,844.43	2
LYCAENIDAE	*Satyrium tetra*	381,267.92	2
HESPERIIDAE	*Cogia outis*	381,267.92	2
HESPERIIDAE	*Amblyscirtes texanae*	383,815.88	2
LYCAENIDAE	*Lycaena gorgon*	384,510.78	2
LYCAENIDAE	*Callophrys johnsoni*	395,860.80	2
HESPERIIDAE	*Poanes aaroni*	397,636.65	2
HESPERIIDAE	*Atrytonopsis python*	405,203.33	2
PIERIDAE	*Colias occidentalis*	412,692.80	2
LYCAENIDAE	*Callophrys hesseli*	413,850.96	2

APPENDIX 7.2 *Continued*

FAMILY	SPECIES	GEOGRAPHICAL RANGE SIZE (KM^2)	AGGREGATION NUMBER
LYCAENIDAE	*Habrodais grunus*	426,976.83	2
HESPERIIDAE	*Hesperia woodgatei*	456,317.01	2
HESPERIIDAE	*Oligoria maculata*	457,089.12	2
LYCAENIDAE	*Callophrys fotis*	467,126.55	2
LYCAENIDAE	*Apodemia nais*	475,774.18	2
HESPERIIDAE	*Amblyscirtes reversa*	497,624.90	2
HESPERIIDAE	*Zestusa dorus*	503,184.09	2
HESPERIIDAE	*Oarisma powesheik*	504,959.94	2
LYCAENIDAE	*Celastrina nigra*	508,588.86	2
HESPERIIDAE	*Euphyes pilatka*	510,827.98	2
LYCAENIDAE	*Calephelis borealis*	512,603.83	2
LYCAENIDAE	*Plebejus lupini*	547,966.47	2
HESPERIIDAE	*Ochlodes yuma*	555,455.93	2
HESPERIIDAE	*Atrytonopsis vierecki*	574,141.00	2
HESPERIIDAE	*Panoquina panoquin*	575,685.22	2
NYMPHALIDAE	*Speyeria nokomis*	595,528.44	2
HESPERIIDAE	*Piruna pirus*	598,230.83	2
HESPERIIDAE	*Amblyscirtes carolina*	607,032.88	2
PIERIDAE	*Anthocharis cethura*	665,636.03	3
HESPERIIDAE	*Euphyes dukesi*	762,458.63	3
NYMPHALIDAE	*Oeneis alberta*	764,852.17	3
LYCAENIDAE	*Calephelis muticum*	806,237.26	3
LYCAENIDAE	*Euphilotes rita*	808,476.38	3
HESPERIIDAE	*Amblyscirtes simius*	810,406.66	3

APPENDIX 7.2 *Continued*

FAMILY	SPECIES	GEOGRAPHICAL RANGE SIZE (KM²)	AGGREGATION NUMBER
NYMPHALIDAE	*Cercyonis meadii*	828,010.76	3
LYCAENIDAE	*Calephelis virginiensis*	829,014.51	3
LYCAENIDAE	*Ministrymon leda*	855,883.94	3
LYCAENIDAE	*Callophrys mossii*	887,849.29	3
PAPILIONIDAE	*Papilio palamedes*	896,651.34	3
LYCAENIDAE	*Satyrium kingi*	918,965.32	3
HESPERIIDAE	*Systasea zampa*	957,339.19	3
LYCAENIDAE	*Satyrium fuliginosum*	961,122.53	3
HESPERIIDAE	*Amblyscirtes alternata*	990,694.34	3
HESPERIIDAE	*Polites sonora*	992,624.62	3
LYCAENIDAE	*Lycaena arota*	1,001,658.30	3
HESPERIIDAE	*Megathymus streckeri*	1,003,820.21	3
HESPERIIDAE	*Hesperia attalus*	1,004,283.48	3
LYCAENIDAE	*Phaeostrymon alcestis*	1,040,109.38	3
HESPERIIDAE	*Hesperia meskei*	1,040,418.23	3
LYCAENIDAE	*Lycaena cupresus*	1,048,525.38	3
PIERIDAE	*Artogeia virginiensis*	1,068,368.61	3
HESPERIIDAE	*Yvretta rhesus*	1,075,085.96	3
HESPERIIDAE	*Poanes massasoit*	1,112,919.35	3
HESPERIIDAE	*Amblyscirtes eos*	1,121,721.41	3
HESPERIIDAE	*Poanes yehl*	1,134,152.38	3
NYMPHALIDAE	*Phyciodes pallida*	1,150,135.06	3
HESPERIIDAE	*Pholisora libya*	1,150,984.38	3

APPENDIX 7.2 *Continued*

FAMILY	SPECIES	GEOGRAPHICAL RANGE SIZE (KM^2)	AGGREGATION NUMBER
NYMPHALIDAE	*Speyeria edwardsii*	1,152,528.60	3
HESPERIIDAE	*Amblyscirtes aenus*	1,169,206.17	3
NYMPHALIDAE	*Speyeria egleis*	1,190,902.46	3
HESPERIIDAE	*Polites draco*	1,238,232.81	3
HESPERIIDAE	*Euphyes conspicua*	1,240,626.35	3
NYMPHALIDAE	*Neonympha areolata*	1,258,076.03	3
LYCAENIDAE	*Lycaena nivalis*	1,264,561.76	3
NYMPHALIDAE	*Boloria epithore*	1,273,827.08	3
NYMPHALIDAE	*Erebia magdalena*	1,312,200.95	3
HESPERIIDAE	*Erynnis telemachus*	1,341,772.76	3
PAPILIONIDAE	*Parnassius clodius*	1,353,045.56	3
HESPERIIDAE	*Erynnis lucilius*	1,353,894.89	3
NYMPHALIDAE	*Lethe portlandia*	1,377,984.72	3
HESPERIIDAE	*Amblyscirtes oslari*	1,399,449.38	3
LYCAENIDAE	*Satyrium californica*	1,411,725.92	3
LYCAENIDAE	*Satyrium behrii*	1,415,895.32	3
HESPERIIDAE	*Hesperopsis alpheus*	1,416,744.64	3
HESPERIIDAE	*Pyrgus ruralis*	1,417,130.69	3
HESPERIIDAE	*Amblyscirtes aesculapius*	1,423,925.26	3
NYMPHALIDAE	*Limenitis lorquini*	1,439,599.10	3
HESPERIIDAE	*Hesperia viridis*	1,443,305.22	3
NYMPHALIDAE	*Chlosyne palla*	1,451,952.86	3
HESPERIIDAE	*Problema byssus*	1,475,347.79	3

APPENDIX 7.2 *Continued*

FAMILY	SPECIES	GEOGRAPHICAL RANGE SIZE (KM^2)	AGGREGATION NUMBER
NYMPHALIDAE	*Lethe creola*	1,477,200.85	3
LYCAENIDAE	*Callophrys irus*	1,506,541.03	3
LYCAENIDAE	*Plebejus shasta*	1,555,570.02	3
NYMPHALIDAE	*Speyeria hydaspe*	1,557,809.14	3
NYMPHALIDAE	*Chlosyne leanira*	1,614,713.64	3
LYCAENIDAE	*Satyrium caryaevorus*	1,617,338.82	3
LYCAENIDAE	*Satyrium saepium*	1,618,574.19	3
HESPERIIDAE	*Hesperia ottoe*	1,626,218.08	3
NYMPHALIDAE	*Cercyonis sthenele*	1,693,237.23	3
HESPERIIDAE	*Hesperia pahaska*	1,697,252.20	3
LYCAENIDAE	*Euphilote enoptes*	1,727,055.65	3
NYMPHALIDAE	*Speyeria coronis*	1,734,004.64	3
HESPERIIDAE	*Thorybes confusis*	1,755,546.51	3
HESPERIIDAE	*Polites sabuleti*	1,755,932.56	3
HESPERIIDAE	*Hesperia sassacus*	1,756,241.41	3
LYCAENIDAE	*Lycaena heteronea*	1,782,801.99	3
PAPILIONIDAE	*Papilio indra*	1,794,229.22	3
LYCAENIDAE	*Callophrys sheridanii*	1,806,042.50	3
LYCAENIDAE	*Euphilotes battoides*	1,816,697.62	3
HESPERIIDAE	*Pyrgus scriptura*	1,817,469.73	3
HESPERIIDAE	*Hesperia nevada*	1,834,610.57	3
NYMPHALIDAE	*Limenitis weidemeyerii*	1,912,593.68	3
LYCAENIDAE	*Lycaena mariposa*	1,923,943.70	3

APPENDIX 7.2 *Continued*

FAMILY	SPECIES	GEOGRAPHICAL RANGE SIZE (KM²)	AGGREGATION NUMBER
LYCAENIDAE	*Calycopis cecrops*	1,926,954.93	3
NYMPHALIDAE	*Chlosyne harrisii*	1,931,896.43	3
HESPERIIDAE	*Atrytone arogos*	1,945,639.99	3
NYMPHALIDAE	*Neominois ridingsii*	1,971,814.52	3
LYCAENIDAE	*Lycaena rubidus*	1,978,145.82	3
NYMPHALIDAE	*Euphydryas editha*	1,981,774.74	3
HESPERIIDAE	*Euphyes bimacula*	2,006,791.10	3
LYCAENIDAE	*Glaucopsyche piasus*	2,029,259.50	3
LYCAENIDAE	*Satyrium sylvinus*	2,081,531.35	3
PAPILIONIDAE	*Papilio eurymedon*	2,081,608.56	3
PIERIDAE	*Pieris chlorodice*	2,093,421.84	3
HESPERIIDAE	*Hesperia juba*	2,110,331.05	3
NYMPHALIDAE	*Oeneis uhleri*	2,137,200.48	3
LYCAENIDAE	*Callophrys affinis*	2,139,439.60	3
PIERIDAE	*Anthocharis midea*	2,146,465.80	3
NYMPHALIDAE	*Speyeria idalia*	2,166,772.29	3
HESPERIIDAE	*Poanes viator*	2,168,934.20	3
PIERIDAE	*Euchloe hyantis*	2,176,269.25	3
PAPILIONIDAE	*Eurytides marcellus*	2,185,225.72	3
NYMPHALIDAE	*Nymphalis californica*	2,209,933.24	3
NYMPHALIDAE	*Oeneis macounii*	2,228,155.04	3
HESPERIIDAE	*Ochlodes sylvanoides*	2,274,404.43	3
HESPERIIDAE	*Hesperia metea*	2,297,799.36	3
PIERIDAE	*Neophasia menapia*	2,364,123.61	3

APPENDIX 7.2 *Continued*

FAMILY	SPECIES	GEOGRAPHICAL RANGE SIZE (KM^2)	AGGREGATION NUMBER
HESPERIIDAE	*Nastra lherminier*	2,390,915.83	3
NYMPHALIDAE	*Lethe appalachia*	2,395,394.06	3
NYMPHALIDAE	*Cercyonis oetus*	2,410,913.48	3
HESPERIIDAE	*Staphylus hayhurstii*	2,428,826.43	3
LYCAENIDAE	*Lycaena epixanthes*	2,469,671.05	3
NYMPHALIDAE	*Speyeria callippe*	2,534,991.55	3
NYMPHALIDAE	*Boloria napaea*	2,545,569.46	3
LYCAENIDAE	*Plebejus icarioides*	2,691,498.25	4
HESPERIIDAE	*Achalarus lyciades*	2,697,983.97	4
NYMPHALIDAE	*Chlosyne gabbii*	2,758,440.19	4
NYMPHALIDAE	*Euphydryas phaeton*	2,772,106.53	4
HESPERIIDAE	*Pompeius verna*	2,812,565.10	4
LYCAENIDAE	*Satyrium edwardsii*	2,840,669.90	4
PAPILIONIDAE	*Papilio troius*	2,849,780.80	4
HESPERIIDAE	*Euphyes dion*	2,911,240.76	4
HESPERIIDAE	*Erynnis baptisiae*	3,022,579.02	5
LYCAENIDAE	*Fixsenia favonius*	3,029,528.01	5
LYCAENIDAE	*Apodemia mormo*	3,034,392.30	5
PIERIDAE	*Euchloe creusa*	3,037,866.80	5
HESPERIIDAE	*Erynnis martialis*	3,039,333.80	5
HESPERIIDAE	*Amblyscirtes hegon*	3,083,189.65	6
PIERIDAE	*Colias pelidne*	3,116,930.86	6
HESPERIIDAE	*Poanes zabulon*	3,189,045.93	6

APPENDIX 7.2 *Continued*

FAMILY	SPECIES	GEOGRAPHICAL RANGE SIZE (KM²)	AGGREGATION NUMBER
PIERIDAE	*Euchloe olympia*	3,197,230.30	6
HESPERIIDAE	*Thorybes bathyllus*	3,251,278.00	6
HESPERIIDAE	*Hesperia leonardus*	3,314,668.23	6
HESPERIIDAE	*Erynnis horatius*	3,347,328.48	6
LYCAENIDAE	*Satyrium acadica*	3,351,034.61	6
PAPILIONIDAE	*Papilio zelicaon*	3,365,859.12	6
NYMPHALIDAE	*Speyeria mormonia*	3,441,989.17	6
PIERIDAE	*Anthocharis sara*	3,475,035.48	6
PAPILIONIDAE	*Parnassius phoebus*	3,490,091.62	6
HESPERIIDAE	*Polites origenes*	3,512,328.39	6
NYMPHALIDAE	*Phyciodes batesii*	3,557,419.61	6
LYCAENIDAE	*Callophrys henrici*	3,620,809.85	6
HESPERIIDAE	*Atrytonopsis hianna*	3,699,951.12	6
LYCAENIDAE	*Plebejus optilete*	3,738,633.83	6
NYMPHALIDAE	*Lethe anthedon*	3,760,175.70	6
LYCAENIDAE	*Lycaena xanthoides*	3,779,246.82	6
NYMPHALIDAE	*Euphydryas chalcedona*	3,814,532.24	6
NYMPHALIDAE	*Erebia epipsodea*	3,858,156.46	6
NYMPHALIDAE	*Lethe eurydice*	3,943,320.19	6
LYCAENIDAE	*Feniseca tarquinius*	3,966,637.91	6
PIERIDAE	*Pieris sisymbrii*	3,979,300.52	6
LYCAENIDAE	*Lycaena dorcas*	4,197,344.38	6
NYMPHALIDAE	*Chlosyne gorgone*	4,217,959.72	6
NYMPHALIDAE	*Polygonia comma*	4,430,212.76	6

APPENDIX 7.2 *Continued*

FAMILY	SPECIES	GEOGRAPHICAL RANGE SIZE (KM²)	AGGREGATION NUMBER
LYCAENIDAE	*Satyrium calanus*	4,434,613.79	6
HESPERIIDAE	*Polites mystic*	4,515,608.12	6
LYCAENIDAE	*Lycaena hyllus*	4,553,441.51	6
HESPERIIDAE	*Poanes hobomok*	4,652,039.96	6
LYCAENIDAE	*Callophrys niphon*	4,689,718.93	6
NYMPHALIDAE	*Megisto cymela*	4,714,580.87	6
PIERIDAE	*Colias alexandra*	4,829,393.63	6
LYCAENIDAE	*Plebejus melissa*	4,838,041.26	6
HESPERIIDAE	*Ancyloxcypha numitor*	4,944,438.02	6
PIERIDAE	*Colias interior*	4,988,448.29	6
PIERIDAE	*Colias scudderii*	5,016,398.67	6
NYMPHALIDAE	*Speyeria aphrodite*	5,018,406.16	6
NYMPHALIDAE	*Oeneis polyxenes*	5,038,017.75	6
PIERIDAE	*Colias nastes*	5,101,562.40	6
NYMPHALIDAE	*Chlosyne nycteis*	5,216,375.16	6
LYCAENIDAE	*Callophrys eryphron*	5,262,624.55	6
LYCAENIDAE	*Callophrys polios*	5,484,451.75	6
PIERIDAE	*Euchloe ausonides*	5,741,873.23	6
NYMPHALIDAE	*Oeneis chryxus*	5,836,919.97	6
LYCAENIDAE	*Satyrium lyparops*	5,841,707.05	6
NYMPHALIDAE	*Speyeria cybele*	5,855,450.61	6
NYMPHALIDAE	*Polygonia satyrus*	6,040,062.11	6
HESPERIIDAE	*Erynnis persius*	6,070,946.51	6

APPENDIX 7.2 *Continued*

FAMILY	SPECIES	GEOGRAPHICAL RANGE SIZE (KM²)	AGGREGATION NUMBER
LYCAENIDAE	*Callophrys gryneus*	6,241,505.61	6
NYMPHALIDAE	*Boloria bellona*	6,382,878.95	6
HESPERIIDAE	*Polites peckius*	6,509,041.72	6
NYMPHALIDAE	*Phyciodes morpheus*	6,637,829.67	6
NYMPHALIDAE	*Polygonia progne*	6,784,453.36	6
HESPERIIDAE	*Epargyreus clarus*	6,798,969.03	6
LYCAENIDAE	*Harkenclenus titus*	6,809,392.51	6
HESPERIIDAE	*Polites themistocles*	6,919,649.82	6
HESPERIIDAE	*Euphyes ruricola*	6,924,900.17	6
PAPILIONIDAE	*Papilio machaon*	7,039,172.45	6
HESPERIIDAE	*Erynnis icelus*	7,145,337.57	6
PIERIDAE	*Pieris callidice*	7,351,645.37	6
HESPERIIDAE	*Amblyscirtes vialis*	7,359,134.83	6
HESPERIIDAE	*Carterocephalus palaemon*	7,433,643.45	6
NYMPHALIDAE	*Speyeria atlantis*	7,475,723.44	6
LYCAENIDAE	*Plebejus idas*	7,659,948.89	6
NYMPHALIDAE	*Boloria eunomia*	7,802,094.34	6
LYCAENIDAE	*Everes amyntula*	7,814,756.94	6
NYMPHALIDAE	*Polygonia faunus*	7,827,805.60	6
LYCAENIDAE	*Callophrys augustus*	7,894,361.48	6
LYCAENIDAE	*Lycaena helloides*	7,961,843.90	6
LYCAENIDAE	*Plebejus saepiolus*	8,076,116.18	6
NYMPHALIDAE	*Polygonia gracilis*	8,097,271.99	6

APPENDIX 7.2 *Continued*

FAMILY	SPECIES	GEOGRAPHICAL RANGE SIZE (KM^2)	AGGREGATION NUMBER
NYMPHALIDAE	*Cercyonis pegala*	8,176,953.74	6
NYMPHALIDAE	*Aglais milberti*	8,342,571.34	6
NYMPHALIDAE	*Boloria selene*	9,092,676.20	6
NYMPHALIDAE	*Limenitis arthemis*	9,094,606.48	6
LYCAENIDAE	*Glaucopsyche lygdamus*	10,210,305.43	7
NYMPHALIDAE	*Coenonympha tullia*	10,403,332.93	7
PAPILIONIDAE	*Papilio glaucus*	12,537,676.60	7

8

DISCONTINUITIES IN URBAN SYSTEMS

Comparison of Regional City-Size Structure in the United States

Ahjond S. Garmestani, Craig R. Allen, and K. Michael Bessey

IN COMPLEX systems, interactions between variables at different scales are not regulated by a central controller (Bak, Tang, and Wiesenfeld 1988; Loreto et al. 1995; Bonabeau 1998). Rather, complex systems organize in a decentralized manner via interactions between agents, variables, and the system itself (Bonabeau 1998). Self-organization manifests in structures that appear at a global scale from interaction between smaller-scale variables (Bonabeau 1998). A self-organized system is characterized by the system's ability to adapt, which leads to broad-scale responses within the system (Hartvigsen, Kinzig, and Peterson 1998). Levin (1998) characterizes the essential elements of a complex system as a sustained diversity and individuality of components, localized interactions among those components, and an autonomous process that selects from among those components. The development of pattern (e.g., aggregations and hierarchical organization) is a consequence of self-organization in a complex system (Levin 1998). As these patterns develop and are expressed, they entrain interactions between variables and agents, which influences future system development (Levin 1998). Complex systems also are conservative. They appear to resist change, or they change slowly despite interchange among and evolution of individual components and the relationships between these components (Levin 1998).

For complex systems such as ecosystems, there has been an intense focus on formalizing scaling laws. West, Brown, and Enquist (1997, 1999) suggest that the scaling of species can be fit to a power law based on the necessity for a space-filling network to supply energy to the cells, invariant capillary size in all vertebrates, and minimal expenditure of energy for the distribution network. The potential then exists that individual allometries of water, nitrogen, and energy flux interact with the fractal scale of habitat to structure the distribution of species (Milne 1997). Scaling laws are useful in studying complex systems because scaling relations may indicate that the system is controlled by a few rules that propagate across a wide range of scales (Meakin 1993; Stanley et al. 1996). Although scaling laws provide powerful tools for studying complex systems, both physics and ecology are just beginning to derive scaling relations based on principles of energy flow and individual behavior (Peters 1983; Milne 1998). Although "a search for universality among systems as disparate as river networks, food webs, oceanic archipelagos, socioeconomic systems, and populations of various taxa could provide a powerful synthesis in ecology" (Milne 1998, 453–454), the utility of scaling laws to describe complex systems is not known, and scaling laws are likely to provide but a partial, simplified picture.

There is growing evidence that ecosystems are governed by processes operating at distinct temporal and spatial scales, which in turn create structure in the landscape with scale-specific pattern (Levin 1992; Milne et al. 1992). Growing evidence demonstrates regular patterns of deviation from scaling laws and from continuous distributions of species (Allen and Holling 2002). Different-size animals in the same ecosystem perceive their environment and operate at different temporal and spatial scales (Peters 1983; Milne, Johnston, and Forman 1989). Holling (1992) suggested that the discontinuous pattern of resource distribution acts on animal communities through species assortment and character displacement, which in turn manifest in a discontinuous distribution of species' body masses. These "discontinuities" are expressed as aggregations and gaps in species' body-mass distributions. For example, rather than a continuous curve of species from smallest to largest, species form several clusters with respect to their body masses (Havlicek and Carpenter 2001). These clusters are separated by distinct breaks, which may indicate the available scales of structure (Raffaelli et al. 2000). The discontinuities between aggregations of species body masses may be indicative of scale breaks, where the scaling laws applicable to larger or smaller species do not apply (Peterson, Allen, and Holling 1998; Allen, Forys, and Holling 1999). Kauffman (1993) proposed that there is increased structural and dynamic complexity at the edge of systems, which in turn makes them highly adaptable. The possibility then exists that this same adaptability and complexity may exist at the edges of structure (e.g., scale breaks) within ecosystems (Allen and Holling 2002). Recent research has revealed that both invasive and extinct species occur at the edges of body-mass aggregations (Allen, Forys, and Holling 1999; Allen 2006). This result has been strengthened by the discovery that birds exhibiting the unique behavior of nomadism are also found at the edges of body-mass aggregations (Allen and Saunders 2002, 2006).

Humans are the dominant species on Earth and are therefore an implicit component of the study of its ecosystems (Vitousek et al. 1997; Grimm et al. 2000). Humans are not disturbances to an ecosystem, but rather drivers and limitations of a particular ecosystem and of the self-organizing biotic and abiotic processes and species therein (Grimm et al. 2000; Allen 2001). As social animals, humans create institutions to regulate knowledge associated with large learning capacities (Pickett et al. 1997). The institutions that govern human population density, population location, and the populations themselves are subject to change through time (Pickett et al. 1997).

An urban system (i.e., a city) is a manifestation of human adaptation to the natural environment (Bessey 2002). Urban systems exhibit spatial patchiness in their social and economic infrastructure (Grimm et al. 2000). Thus, urban systems, much like ecosystems, are subject to a hierarchy of structure and processes that govern the function and growth of cities at a variety of scales (Marshall 1989; Grimm et al. 2000; Bessey 2002). City size is constrained by numerous factors, including the hydrologic infrastructure (e.g., canals, pipes, and storm drains), land, development opportunities, and access to transportation (Spirn 1984; Grimm et al. 2000). The interaction between cities and natural processes affect cities' economic and health conditions (Spirn 1984). However, the type of processes in urban systems (e.g., human-induced changes as well as political-economic, institutional, and cultural forces) probably differs from the makeup of the processes structuring ecological communities (Dow 2000).

Urban systems display a diverse spatial mosaic that warrants study of their structure and process (Pickett et al. 2001). For example, urban systems' spatial heterogeneity is typically established and maintained by government (e.g., zoning regulations enforced by zoning boards and courts) and influenced on a different scale by other institutions such as businesses and community associations (Grimm et al. 2000). The processes operating at different scales in ecological systems mirror those that operate in urban systems with respect to the structuring of the systems (Pickett et al. 1997; Dow 2000). For example, a variable that has an effect at a local level, such as movement of businesses or national policy, may have derived from a different scale (Dow 2000).

Bessey (2002) has suggested that these processes are corollaries of the "slaving principle," in which large-scale, slow processes (e.g., national economies) enslave small-scale, fast processes (e.g., regional and city economies). Pattern is a function of process (Solé and Manrubia 1995). Thus, the signature that these processes impart on the landscape (e.g., cities as well as their size and distribution) may illuminate the nature of these hierarchical processes within complex systems (Bessey 2002). For example, urban primacy and the clustering of regional city-size distributions suggest spatial and temporal discontinuity in urban systems (Bessey 2002). Thus, city-size distributions may offer information about complex evolving systems that would not be subject to the same processes as in ecosystems, but may reflect other fundamental processes that manifest in discontinuities (Bessey 2002).

Much as Holling (1992) suggested for ecosystems, the physical structure of the environment plays a crucial role in shaping the landscape of an urban system (Dow 2000). For example, canals, railways, and roads structure in part the flow of commerce and people in and out of cities. Also, in the structuring of urban systems, persons of wealth will locate their neighborhoods at higher elevations, which reflects historical patterns of belief about health and disease (Meyer 1994; Dow 2000). Variables such as wealth, education, status, property, and power, distributed inequitably, are expressed at different spatial and temporal scales and add to the hierarchical structuring of urban systems (Pickett et al. 1997; Pickett et al. 2001). Similar to hierarchical structure in ecosystems, the spatial heterogeneity in urban systems is affected by the generation, flow, and concentration of resources (Pickett et al. 1997). Further, an urban system's resilience, much like an ecosystem's, buffers the system against disturbances, whether they are planned (e.g., smart growth) or unplanned (e.g., the collapse of the Silicon Valley "Dotcoms" in 2000; Bessey 2002).

Urban distributions have been described by Zipf's law, or the rank-size rule (Zipf 1949). Gabaix (1999) states that Zipf's law for cities is an empirical fact in economics and in the social sciences in general. Zipf's law predicts that city-size distributions will have a continuous distribution and conform to the restraints of a linear power law (Gabaix 1999). If an urban system develops under these power laws, the resulting steady-state distribution of city sizes will approximate a rank-size distribution (Simon 1955). Supporters of the proposition that urban distributions conform to Zipf's law believe that this fractal scaling distribution describes urban systems structured by a hierarchy of time-minimizing spatial constraints (Zipf 1949). This rank-size relationship for urban systems, as described by Zipf's law, is believed to be a reflection of a steady-state condition (Gabaix 1999). Thus, the assumption is that city sizes of a certain range will have similar growth processes (Gibrat's law) regardless of the particulars driving the growth of cities and that these cities' distribution will conform to Zipf's law (Gibrat 1957; Gabaix 1999). Further, it is assumed that there is a constant rate of new cities competing with established cities for resources (Yule, Stuart, and Kendall 1971).

City sizes are thought to conform to a power law (Zipf's law) due to the invariance of growth processes at the range of possible scales (Gabaix 1999). However, urban systems are not deterministic (Pickett et al. 2001). Rather, they are entrained by stochastic, historical, and hierarchical influences that make their development different from predictions based on physical laws (Pickett et al. 2001). Further, city sizes are defined by the maximum potential welfare of the participants in the economy, and these participants operate at different scales (Henderson 1974; Kline, Moses, and Alig 2001). Gabaix (1999) has also intimated that scale-specific processes are at work on city size: for cities larger than a certain size, shocks (e.g., policy or natural disasters) stop declining with the size of the cities in question. In addition, Lynch (1960) identified five spatial scales for urban systems, including district, edge, path, node, and landmark. These spatial scales manifest as neighborhoods, commercial-residential divides, and transportation corridors (Dow 2000). Gabaix (1999) contends that Zipf's law is not disproved even if two cities in the rank order are quite close in size. However, the inherent deficiency of Zipf's law is that it does not capture evidence of hierarchical structure in urban distributions (Bessey 2002). Deviations from Zipf's law may provide an additional source of information about the state of the system and a starting point in the search for reasons for such deviations (Dziewonski 1972). As Gabaix (1999) has acknowledged, if city sizes are indeed structured by a hierarchy of processes operating at different scales, then a power law probably does not capture the actual structure in urban systems.

Bessey (2002) found that bimodality and polymodality are defining features of U.S. urban systems at national and regional scales. He utilized rank-size and constant-Gini models to analyze national and regional city-size data. The models revealed that there were departures from the Zipf prediction and increasing population concentration in the largest cities (i.e., upper tail of the city-size distribution) in each region. At a finer scale, individual cities often followed paths that were sharply discontinuous in their growth trajectories (Bessey 2002). For individual cities, Bessey found that there were periods of static behavior linked by periods of oscillatory turbulence or instability, constrained by regional and national processes. In addition, he identified at a regional level that some cities' tenure within a particular mode was sometimes highly transient. For example, in the Rocky Mountain region of the United States, Ogden, Utah, had rates of growth at 43%, 37%, 27%, and 26% over the past four decades, respectively, which resulted in a discontinuous growth process (Bessey 2002).

Discontinuous patterns have been found in international economic data (Summers and Heston 1991), where the variable of interest has been gross domestic product (GDP) per capita, measured for 120 countries over a thirty-year time frame. A discontinuous distribution was found to persist over time, and the overall structure bound the trajectories of growth for individual countries within the data set. The overall structure persisted through the time series, raising the question regarding the role of history versus the role of structuring forces in determining the pattern. Barro (1997), for example, has hypothesized the existence of a discrete number of "convergence clubs" in per capita GDP data—that is, sets of countries whose similar attributes "entrain" their economic performance—but does not indicate what forces would cause clustering in the determining attributes. Further tests of the convergence and convergence-club hypotheses have been performed using economic data from other scales, including U.S. states (Barro and Salai-i-Martin 1992; Kenworthy 1999) and U.S. counties, but the conclusions about trends in inequality indices are mixed.

We hypothesize that hierarchical structure is strongly self-organizing and conservative in urban systems. If this is so, we expect discontinuities to persist despite the normal dynamics of the system. Individual cities may prosper or perish, but, overall, the hierarchical structure should be little affected unless the system is pushed beyond the limits of its resilience. In that case, we should observe a loss of hierarchical structure (at some scales) during the system collapse and a reemergence of it during the subsequent reorganization (Holling 2001;

Gunderson and Holling 2002). We test these predictions here with empirical data sets that reflect system structure over time.

METHODS

We define an *urban system* as a human settlement greater than some population size that satisfies the functional requirements of that population (Bessey 2002). The cutoff for determining what is urban is arbitrary and arises from practical rather than theoretical considerations (Marshall 1989). For example, Boston had a 1990 population of 574,000 or 2.8 million or 3.2 million or 5.8 million depending on whether the defining reference is the city, the urbanized area, the primary metropolitan statistical area, or the consolidated metropolitan statistical area (Bessey 2002). A precise definition of city size in a distribution of urban systems is critical because the manner in which these city sizes are defined may account for much of the variation between Zipf's law and empirical data (Berry 1971). Our analysis used a U.S. Census data set that incorporates the definition of the urban area as its foundation, as described by Bessey (2000). An urbanized area comprises a central place and the urban fringe, which includes other "places" (Bessey 2002). The U.S. Bureau of the Census officially defines a *place* as a concentration of population that must have a name and be locally recognized, although it may or may not be legally incorporated under its state's laws (Bessey 2002). Outside of urban areas, an *urban place* is any incorporated place or census-designated place (unincorporated) with at least 2,500 inhabitants (Bessey 2002). U.S. Bureau of Economic Analysis (BEA) regions were used as the units of analysis in this study (Bessey 2002).

Many of the census units have evolved through several definitional changes over the past 120 years (Bessey 2002). BEA regions comprise defined entities whose boundaries hold historically, and we aggregated the urban-area data into regional units (Bessey 2002). Analyzing the data based on BEA regions allowed for analysis along smaller and more uniform biophysical, economic, and sociocultural characteristics, but also along primarily functional attributes such as commute-to-work patterns and newspaper circulation (Bessey 2002). With time, investments in the built environment, in particular transportation and communication infrastructures, might overcome biophysical and geographic barriers, resulting in functional integration between cities and regions (Bessey 2002). Initial conditions (geophysical and economic) can loom large in competitive city-growth processes (Bessey 2002). Dendrinos (1992) describes the existence of a relative, per capita, product developmental threshold below which urban wealth variations over time are almost negligible. A city's relative population share and wealth appear to depend heavily on its past and current location relative to this threshold (Bessey 2002). Bessey (2002) states that capturing the essence of comparative advantages within an environment of interacting locations and spatial agglomeration gradients reveals the workings of a governing principle controlling the flow of stocks in and out of individual urban areas.

We ranked urban systems in order of population size to determine whether discontinuities existed within the city-size distributions. We used BEA data sets of cities in the southwestern and southeastern regions of the United States. We chose the regional analysis because there is landscape self-similarity of processes in the regions of analysis and because national-level analyses follow geopolitical boundaries that obscure regional pattern. We analyzed city-size distributions by using simulations that compared actual data with a null distribution established by estimating a continuous unimodal kernel distribution of the log-transformed data (Silverman 1981). We determined the significance of discontinuities in the data by calculating the probability that the observed discontinuities were chance events and by comparing observed values with the output of one thousand simulations from the null set (Restrepo, Renjifo, and Marples 1997). Because *n* in our eleven data sets varied from 48 cities in 1890 to 161 cities in 1990 for the southwestern region and from 50 cities in 1860 to 310 cities in 1990 for the southeastern region, we maintained a constant statistical power of approximately 0.50 for detecting discontinuities (Lipsey 1990). Although we believe that the application of a null model is the best method for determining city-size aggregations, we confirmed our results with a form of the split moving-window boundary analysis (Webster 1978; Ludwig and Cornelius 1987) and hierarchical cluster analysis (Ward's option; SAS Institute 1999). A *gap (discontinuity)* was defined as an area between successive city sizes that significantly exceeded the discontinuities generated by the continuous null distribution (Restrepo, Renjifo, and Marples 1997; Allen, Forys, and Holling 1999). An *aggregation* was defined as a grouping of three or more cities with populations not exceeding the expectation of the null distribution (Allen, Forys, and Holling 1999). City-size aggregations were defined by the two end-point cities that indicated either the upper extreme or the lower extreme of the aggregation (Allen, Forys, and Holling 1999).

RESULTS

For the southwestern region of the United States, there were 48 cities in 1890 and 161 cities in 1990 (table 8.1). Within each decade, city sizes ranged from 2,541 to 38,067 people in 1890 and from 10,030 to 3,198,259 people in 1990 (table 8.1). Beginning in 1890, the largest city in the southwestern region of the United States was Dallas (table 8.2). For the next three decades (1900–1920), San Antonio was the largest city in the region, and then Houston from 1930 to 1970 (table 8.2). Finally, from 1980 to 1990, Dallas–Fort Worth reascended to the position of largest city in the region after Dallas and Fort Worth merged into one urbanized

TABLE 8.1 Summary of City Data by Decade for the Southwestern Region of the United States

YEAR	LARGEST CITY	LARGEST CITY POPULATION	NUMBER OF DISCONTINUITIES	NUMBER OF CITIES
1890	Dallas	38,067	6	48
1900	San Antonio	54,000	7	54
1910	San Antonio	99,000	6	53
1920	San Antonio	68,700	5	55
1930	Houston	295,700	6	73
1940	Houston	416,100	5	69
1950	Houston	701,600	5	94
1960	Houston	1,140,000	5	120
1970	Houston	1,677,863	4	123
1980	Dallas–Fort Worth	2,451,390	4	149
1990	Dallas–Fort Worth	3,198,259	6	161

region (table 8.2). These three cities represent the dominant cities of this region and have jockeyed for position of largest city over the course of the past century (table 8.2).

The decadal city-size distributions for the southwestern region of the United States were discontinuous (figs. 8.1 to 8.11). We identified distinct aggregations of cities in each decade by all methods of analysis. We observed four to seven discontinuities in each decadal data set (table 8.1). This structure is significant because random draws of the same *n* from the null model revealed that 91% of the outputs randomly generated were either unimodal or bimodal in their distribution, and fewer than 1% had more than four discontinuities (see chap. 5 in this volume). Further, despite great change in the number of cities across decades, the dominance of a few cities and the number of aggregations remained consistent over time. For each time period analyzed, there is a range of city sizes, a different number of cities represented, and a different hierarchical relationship of the cities.

There is no evidence of urban primacy for the southwestern region of the United States, as has been suggested by Jefferson (1939). It is interesting to track the movement of Galveston, Houston, and Phoenix, in particular, to demonstrate

TABLE 8.2 Cities in the Aggregation at the Upper Tail of the Southwestern U.S. Region Distribution by Decade

1890	1900	1910	1920	1930	1940
Dallas	San Antonio	San Antonio	San Antonio	Houston	Houston
San Antonio	Houston	Dallas	Dallas	Dallas	Dallas
Houston	Dallas	Houston	Houston	San Antonio	San Antonio
Galveston	Galveston	Fort Worth	Fort Worth		
Fort Worth		Oklahoma City	Oklahoma City		
Austin			El Paso		
Waco			Tulsa, Okla.		
1950	**1960**	**1970**	**1980**	**1990**	
Houston	Houston	Houston	Dallas–Fort Worth	Dallas–Fort Worth	
Dallas	Dallas	Dallas	Houston	Houston	
San Antonio	San Antonio	Phoenix, Ariz.	Phoenix, Ariz.	Phoenix, Ariz.	
	Phoenix, Ariz.	San Antonio	San Antonio	San Antonio	
	Fort Worth	Fort Worth	Oklahoma City	Oklahoma City	
	Oklahoma City	Oklahoma City			

change over time in the rank of cities. In 1890, Galveston (29,084) and Houston (27,557) had comparable populations and were members of the second-largest aggregation of cities.

Phoenix (3,152), however, was a small town, and a member of a large aggregation with numerous cities of similar small size. By 1900, Houston ascended to the top aggregation, but Galveston descended from the third-ranked city in the region in 1890 to the seventh-ranked city in 1910. This trend continued as Galveston continued a slow slide until it settled into a midrange aggregation by 1990 with a population of 58,263. By 1900, Phoenix moved into a midrange aggregation with a population of 5,544, and moved slightly up in 1910 with a population of 11,134. By 1930, Phoenix had grown to 67,100 people and was the eighth-largest city in the region, surpassing Galveston. By 1960, Phoenix was the fourth-largest city in the region with a population of 552,043. By 1970, it ascended to the third-largest city in the region, where it remained as of 1990, with a population of 2,006,239.

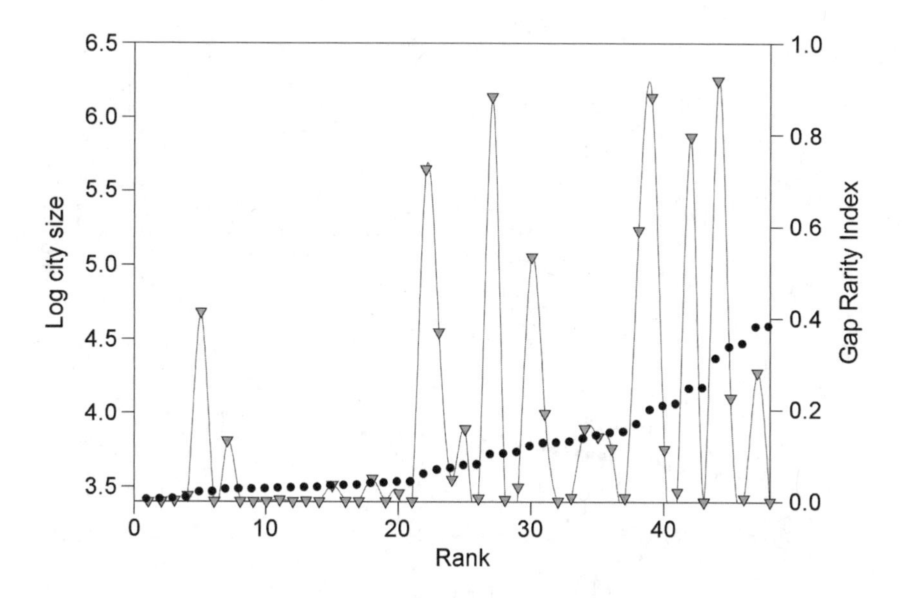

FIGURE 8.1 Log-transformed distribution of city sizes overlaid with Gap Rarity Index values for 1890 U.S. Census data.

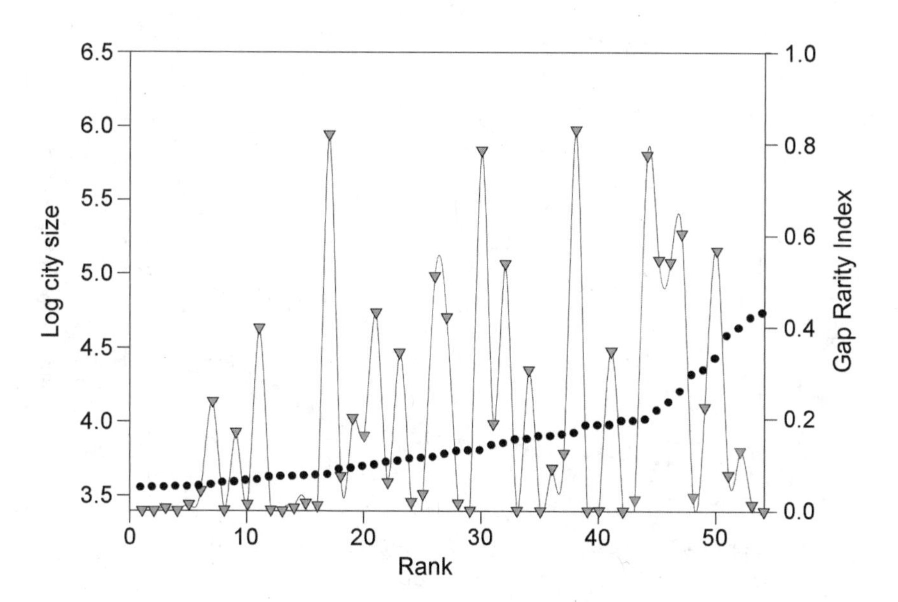

FIGURE 8.2 Log-transformed distribution of city sizes overlaid with Gap Rarity Index values for 1900 U.S. Census data.

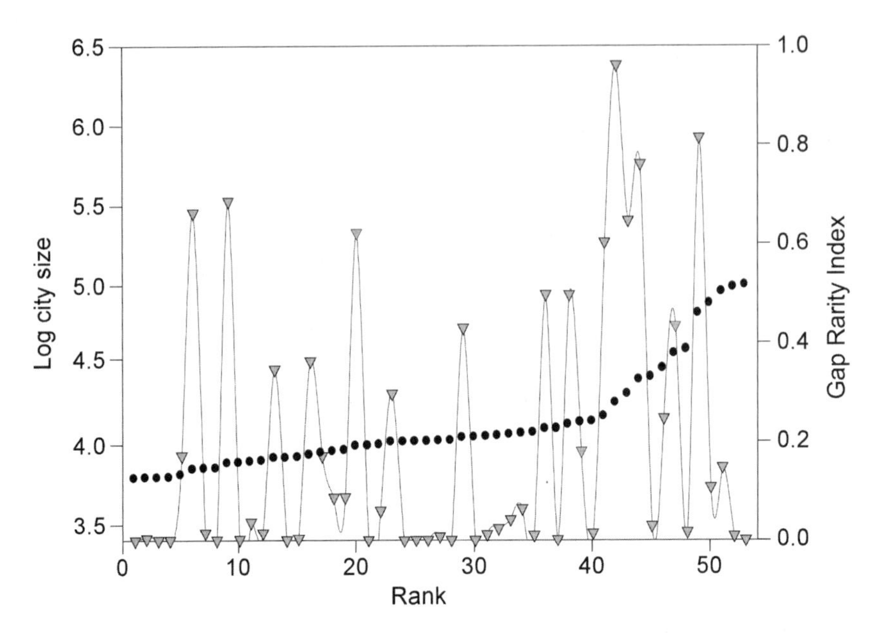

FIGURE 8.3 Log-transformed distribution of city sizes overlaid with Gap Rarity Index values for 1910 U.S. Census data.

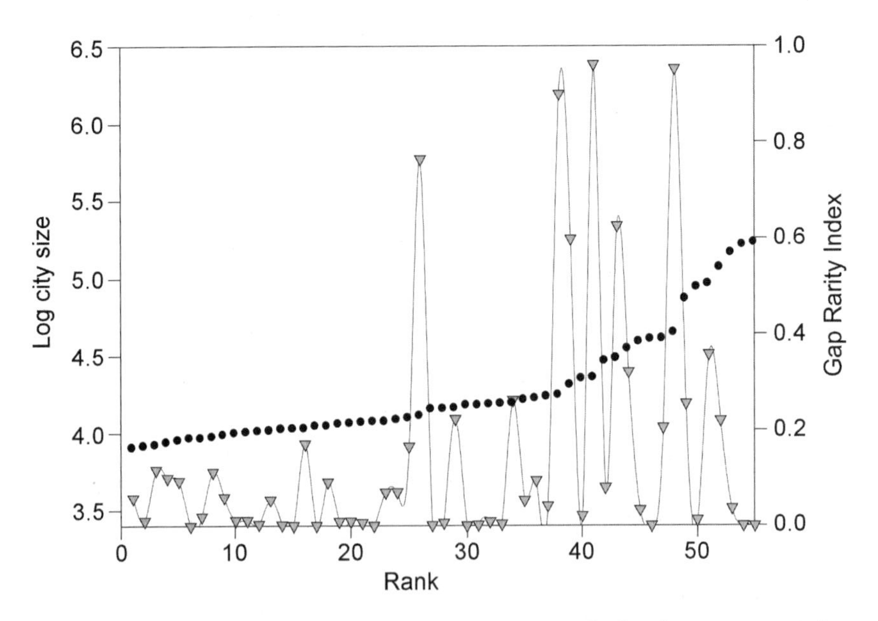

FIGURE 8.4 Log-transformed distribution of city sizes overlaid with Gap Rarity Index values for 1920 U.S. Census data.

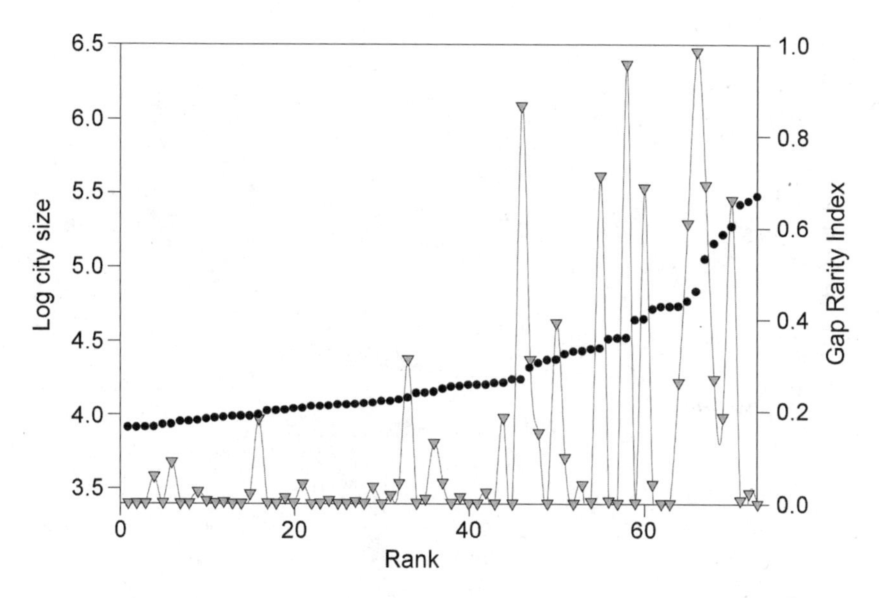

FIGURE 8.5 Log-transformed distribution of city sizes overlaid with Gap Rarity Index values for 1930 U.S. Census data.

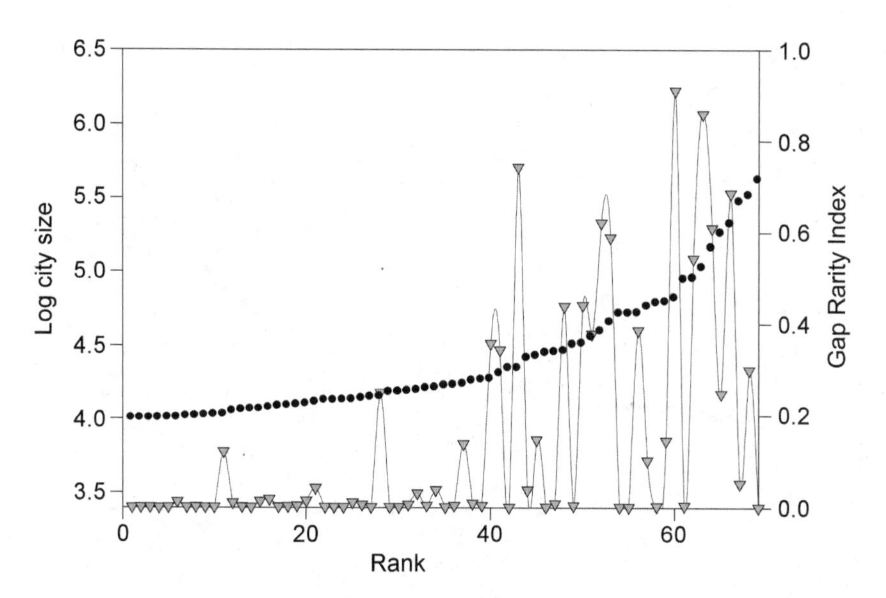

FIGURE 8.6 Log-transformed distribution of city sizes overlaid with Gap Rarity Index values for 1940 U.S. Census data.

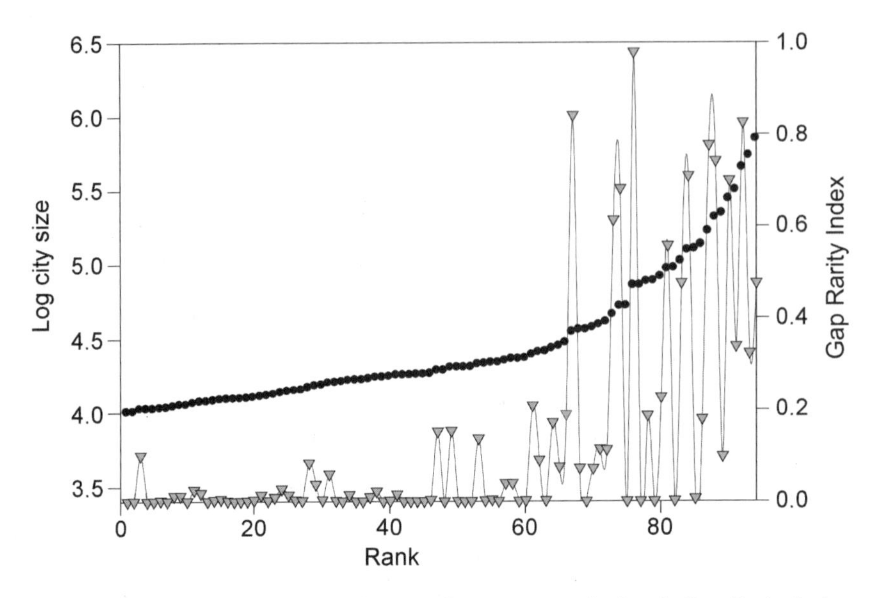

FIGURE 8.7 Log-transformed distribution of city sizes overlaid with Gap Rarity Index values for 1950 U.S. Census data.

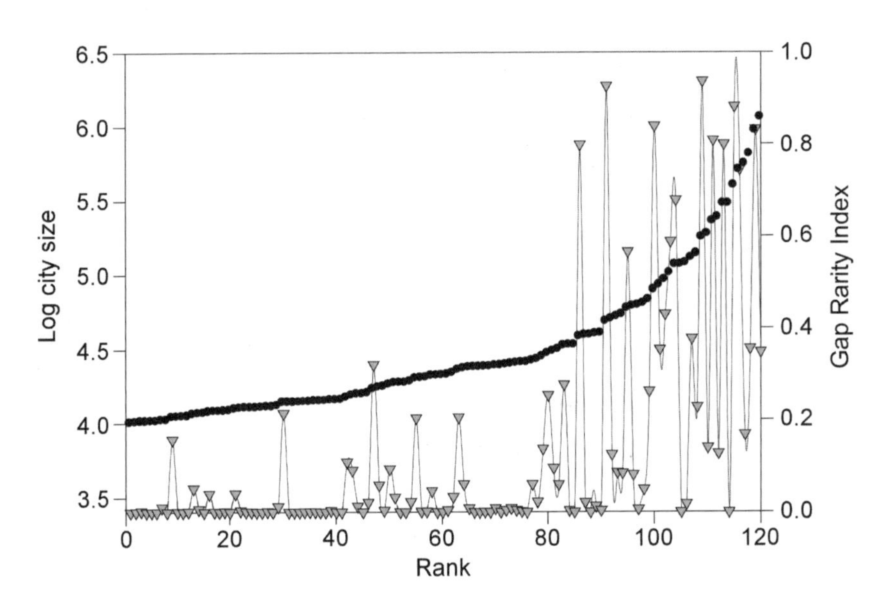

FIGURE 8.8 Log-transformed distribution of city sizes overlaid with Gap Rarity Index values for 1960 U.S. Census data.

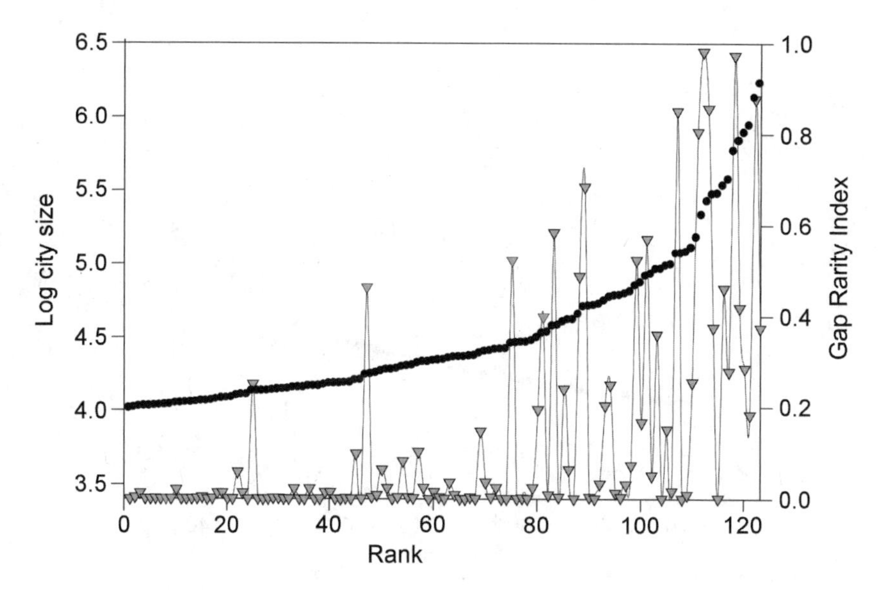

FIGURE 8.9 Log-transformed distribution of city sizes overlaid with Gap Rarity Index values for 1970 U.S. Census data.

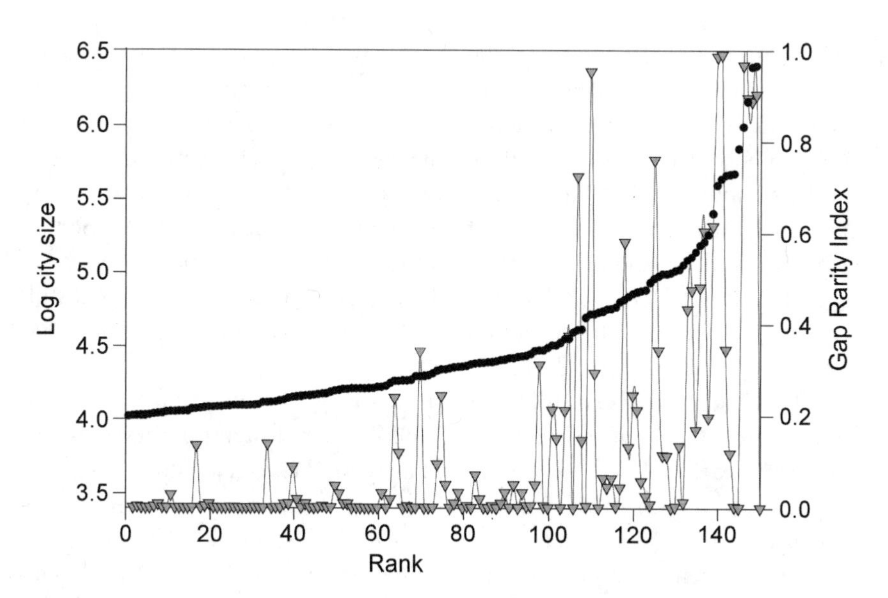

FIGURE 8.10 Log-transformed distribution of city sizes overlaid with Gap Rarity Index values for 1980 U.S. Census data.

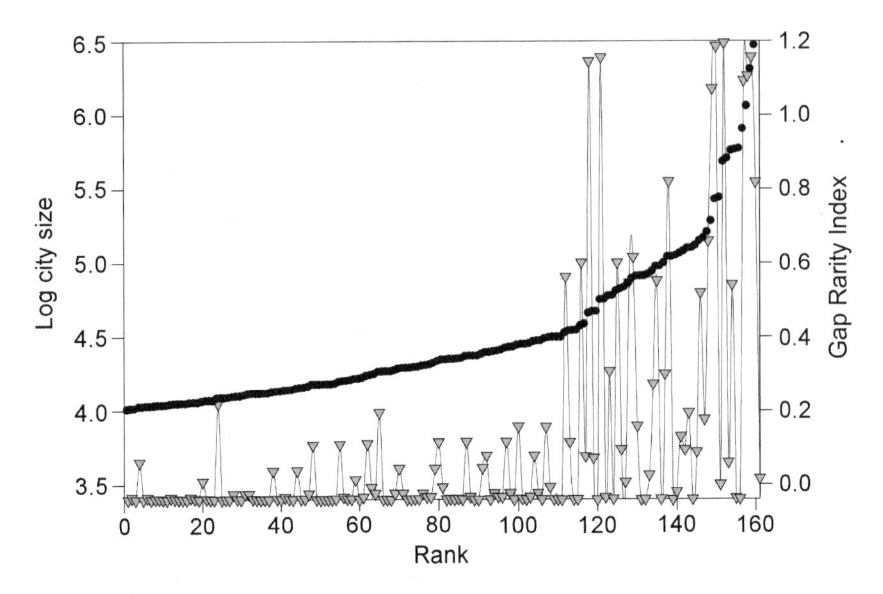

FIGURE 8.11 Log-transformed distribution of city sizes overlaid with Gap Rarity Index values for 1990 U.S. Census data.

For the southeastern region of the United States, there were 50 cities in 1860 and 310 cities in 1990 (table 8.3). City sizes ranged from 2,546 to 181,000 in 1860 and from 10,005 to 2,157,806 in 1990 (table 8.3). Beginning in 1860 and continuing until 1950, the largest city in the southeastern region of the United States was New Orleans (table 8.3). From 1960 to 1970, Miami-Hialeah was the largest city in the region, and then Atlanta was from 1980 to 1990 (table 8.3). From 1860 to 1930, New Orleans and Louisville, Kentucky, were the two dominant cities of the region. After 1930, Louisville's clout in the region began to ebb, although it still remained a member of the aggregation at the upper tail of the city-size distribution for each decade in the data set.

The decadal city-size distributions for the southeastern region of the United States were discontinuous (figs. 8.12 to 8.25). We identified distinct aggregations of cities in each decade by all methods of analysis. We observed three to six discontinuities in each decadal dataset (table 8.3). Despite great change in the number of cities across decades, the dominance of a few cities and the number of aggregations remained consistent over time. For each time period analyzed, there is a range of city sizes, a different number of cities represented, and a different hierarchical relationship among the cities.

There is no evidence of urban primacy for the southeastern region of the United States, as has been suggested by Jefferson (1939). From 1860 to 1900, New Orleans and Louisville were the two dominant cities of the region. Atlanta was

TABLE 8.3 Summary of City Data by Decade for the Southeastern Region of the United States

YEAR	LARGEST CITY	LARGEST CITY POPULATION	NUMBER OF DISCONTINUITIES	NUMBER OF CITIES
1860	New Orleans	181,000	4	50
1870	New Orleans	193,000	4	70
1880	New Orleans	220,000	4	54
1890	New Orleans	248,000	3	57
1900	New Orleans	290,000	3	61
1910	New Orleans	345,000	5	71
1920	New Orleans	389,400	6	97
1930	New Orleans	486,500	4	136
1940	New Orleans	530,600	5	156
1950	New Orleans	649,800	5	189
1960	Miami-Hialeah	852,705	5	249
1970	Miami-Hialeah	1,219,661	3	287
1980	Atlanta	1,613,357	4	306
1990	Atlanta	2,157,806	3	310

the fourth-largest city in the region and had begun its gradual ascent to the position of the dominant city in the region, but Wheeling, West Virginia, also was a member of the aggregation of the largest cities at the upper tail of the distribution and was the ninth-largest city in the region in 1900. Over the course of the twentieth century, Atlanta grew into the urban hub of the southeastern region, whereas Wheeling continued a slow descent until it settled as the seventy-second-largest city in 1990 and a member of the large aggregation of small cities in the lower tail of the city-size distribution.

From 1950 to 1990, the state of Florida became a focal point of growth not only in the southeastern region, but in the United States as a whole. By 1950, Miami was the fourth-largest city, Jacksonville was the tenth-largest city, and Tampa was the twelfth-largest city in the region. Thereafter, Florida cities increased in population and became some of the largest cities in the region. By 1990, Miami was the

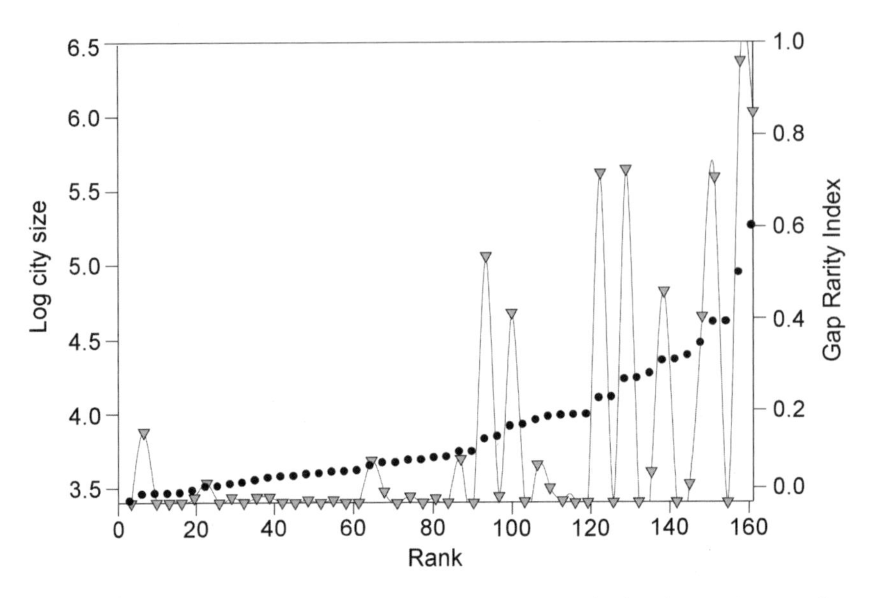

FIGURE 8.12 Log-transformed distribution of city sizes overlaid with Gap Rarity Index values for 1860 U.S. Census data.

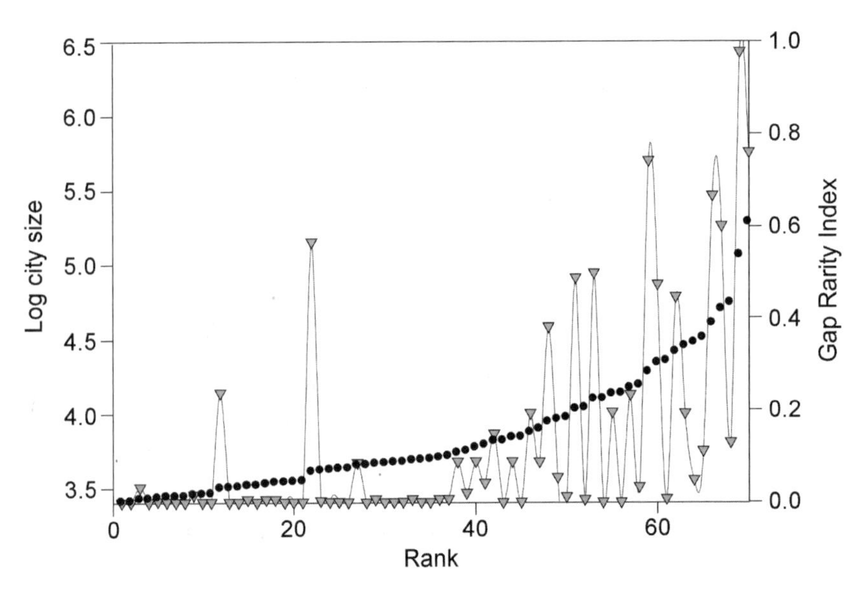

FIGURE 8.13 Log-transformed distribution of city sizes overlaid with Gap Rarity Index values for 1870 U.S. Census data.

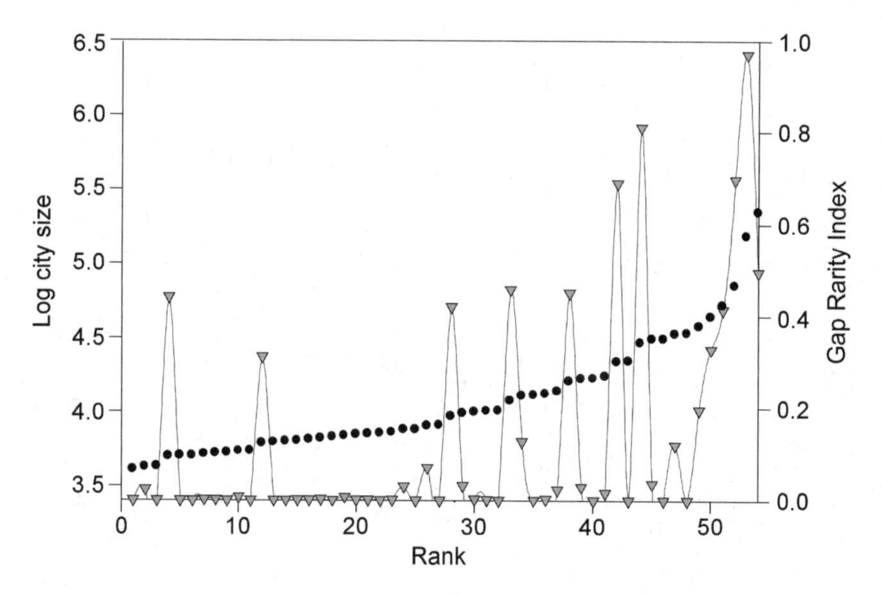

FIGURE 8.14 Log-transformed distribution of city sizes overlaid with Gap Rarity Index values for 1880 U.S. Census data.

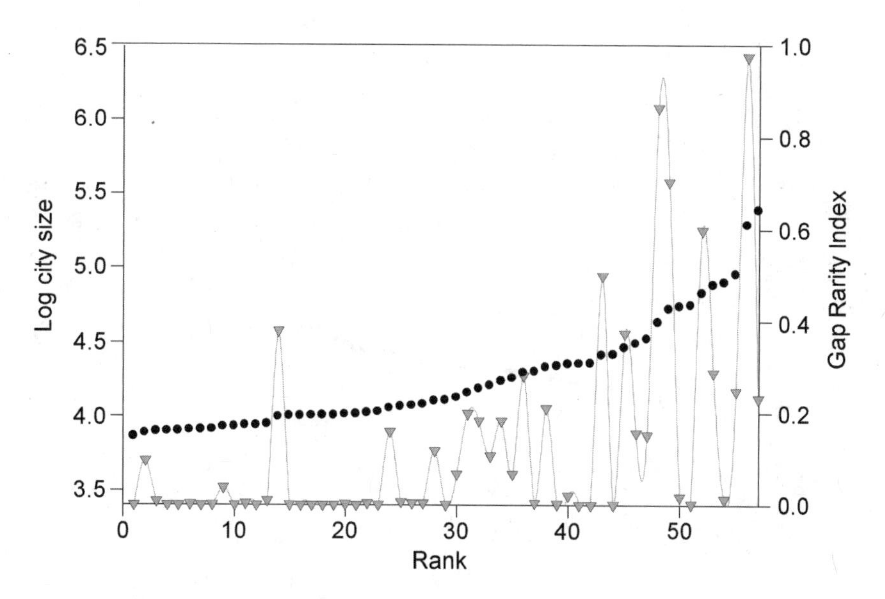

FIGURE 8.15 Log-transformed distribution of city sizes overlaid with Gap Rarity Index values for 1890 U.S. Census data.

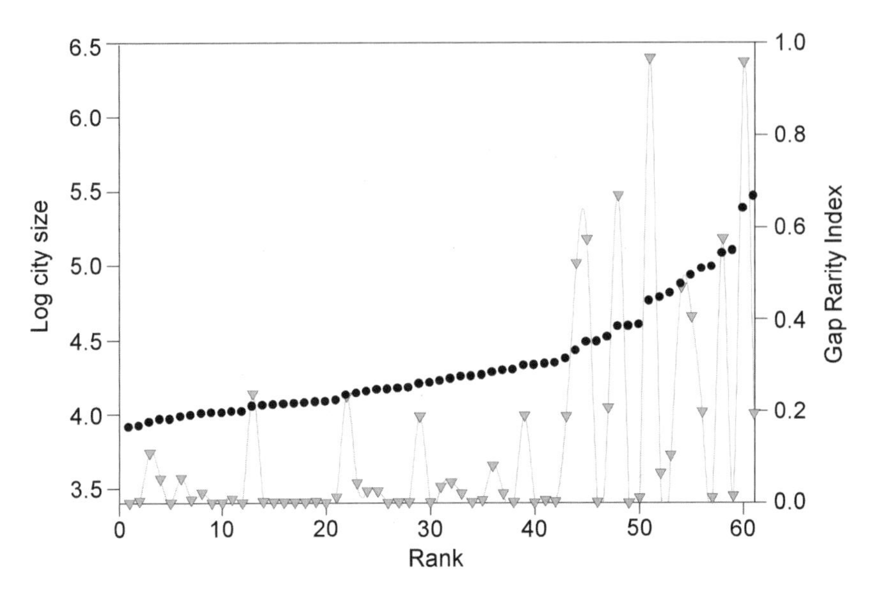

FIGURE 8.16 Log-transformed distribution of city sizes overlaid with Gap Rarity Index values for 1900 U.S. Census data.

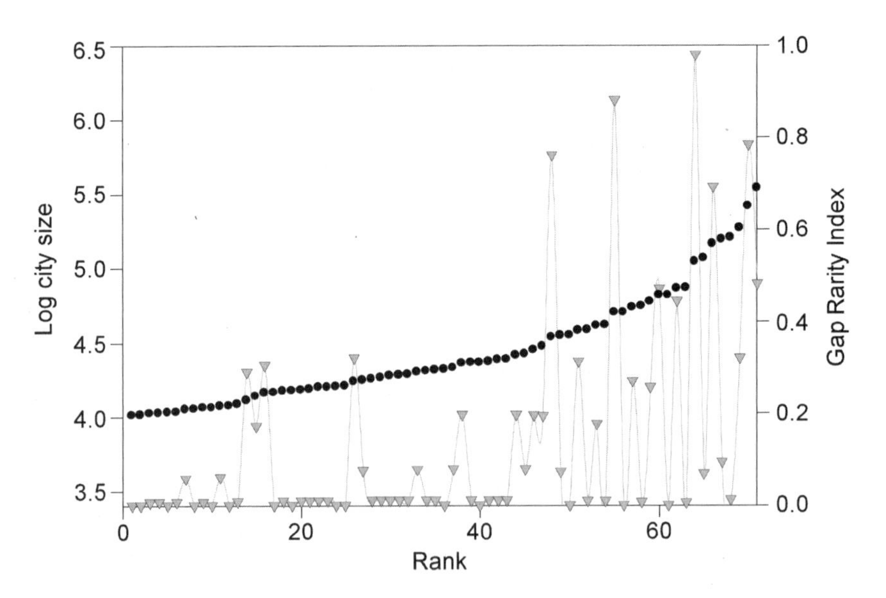

FIGURE 8.17 Log-transformed distribution of city sizes overlaid with Gap Rarity Index values for 1910 U.S. Census data.

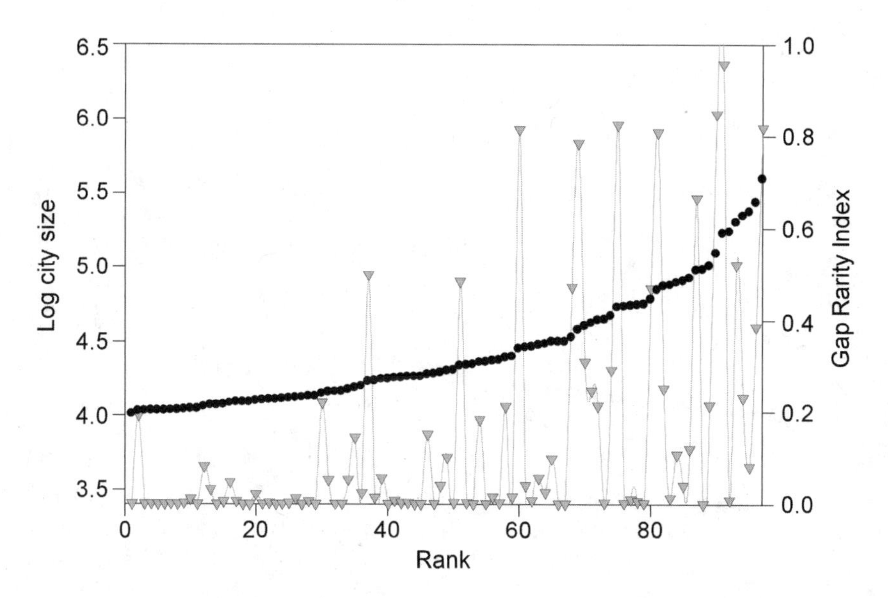

FIGURE 8.18 Log-transformed distribution of city sizes overlaid with Gap Rarity Index values for 1920 U.S. Census data.

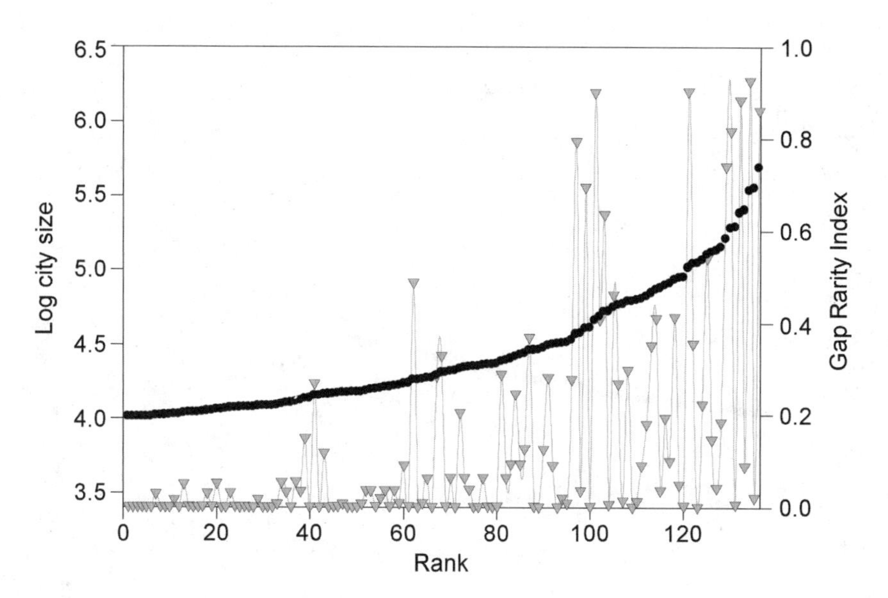

FIGURE 8.19 Log-transformed distribution of city sizes overlaid with Gap Rarity Index values for 1930 U.S. Census data.

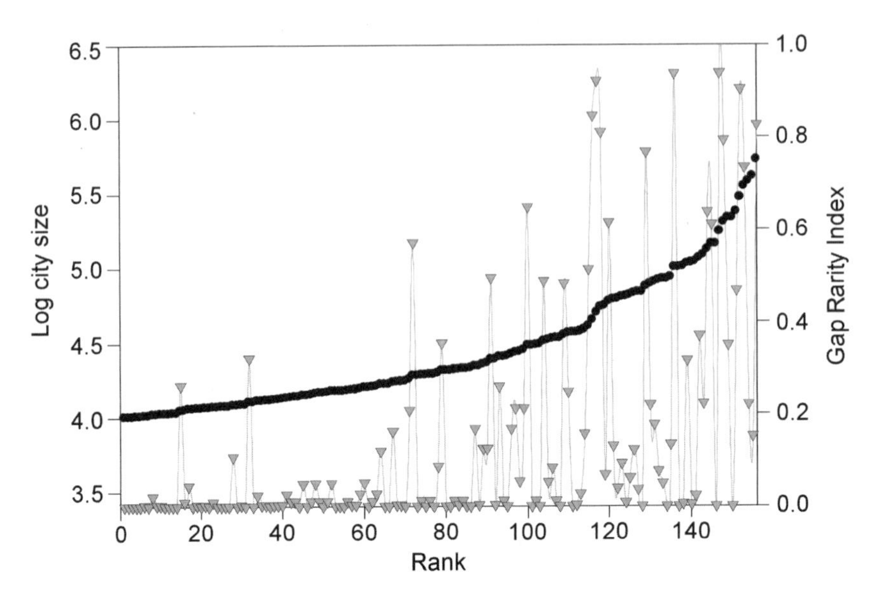

FIGURE 8.20 Log-transformed distribution of city sizes overlaid with Gap Rarity Index values for 1940 U.S. Census data.

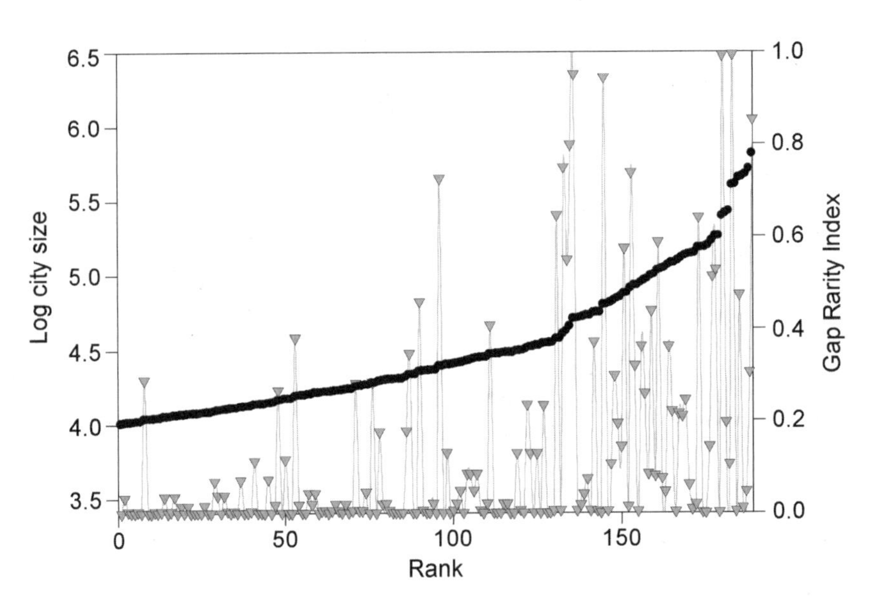

FIGURE 8.21 Log-transformed distribution of city sizes overlaid with Gap Rarity Index values for 1950 U.S. Census data.

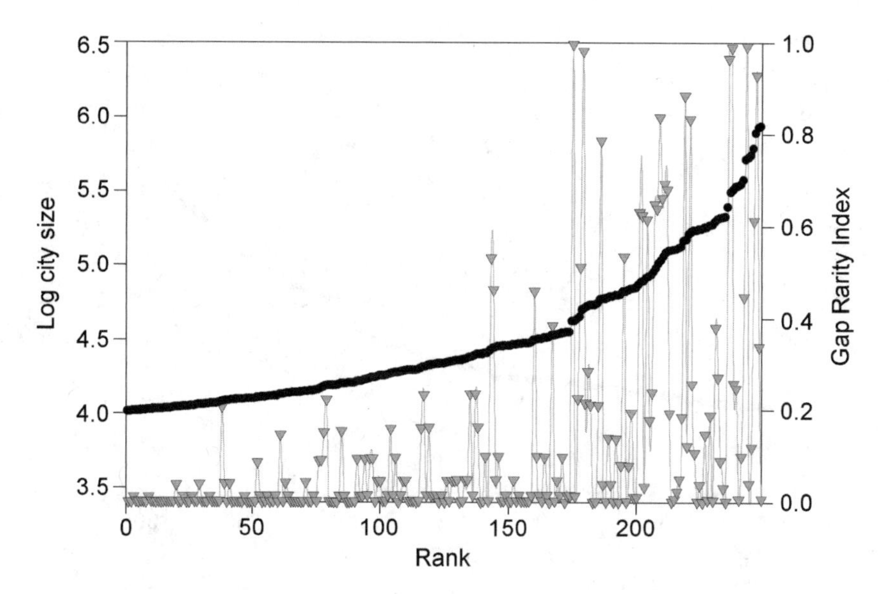

FIGURE 8.22 Log-transformed distribution of city sizes overlaid with Gap Rarity Index values for 1960 U.S. Census data.

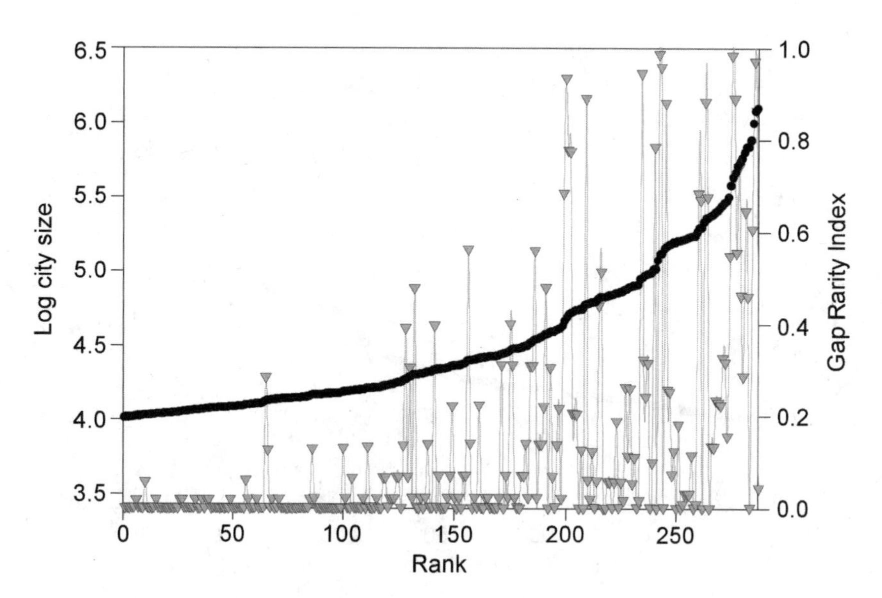

FIGURE 8.23 Log-transformed distribution of city sizes overlaid with Gap Rarity Index values for 1970 U.S. Census data.

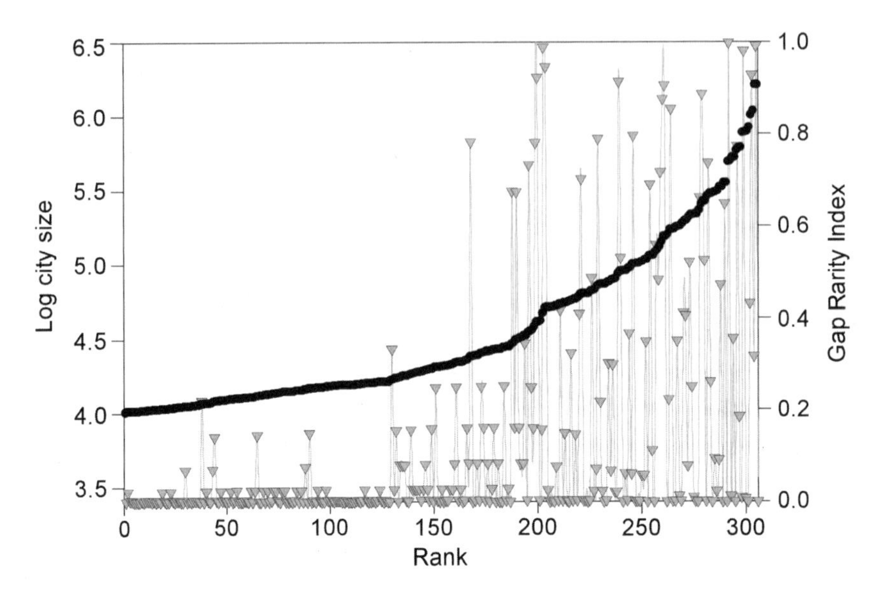

FIGURE 8.24 Log-transformed distribution of city sizes overlaid with Gap Rarity Index values for 1980 U.S. Census data.

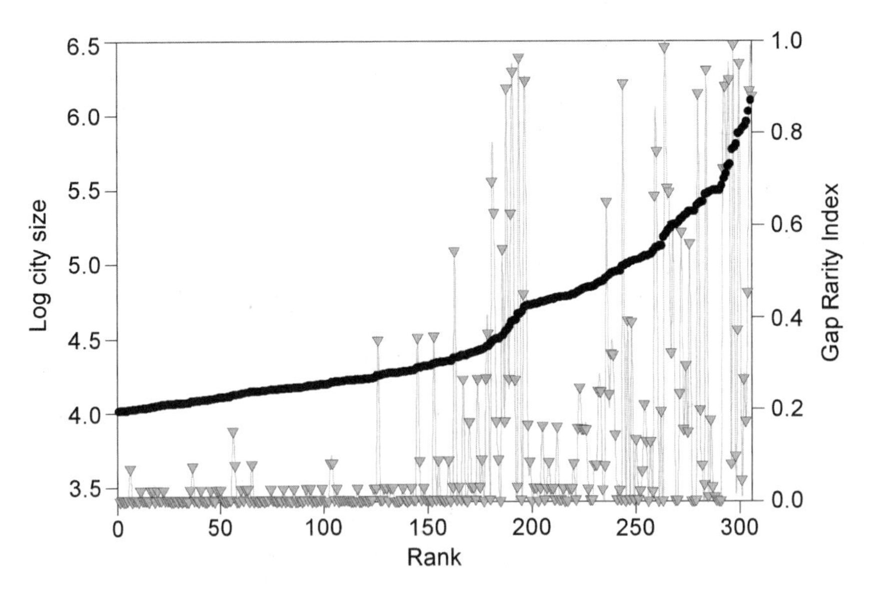

FIGURE 8.25 Log-transformed distribution of city sizes overlaid with Gap Rarity Index values for 1990 U.S. Census data.

second-largest city in the region, and Tampa, Fort Lauderdale, Orlando, West Palm Beach, and Jacksonville were members of the aggregation of the largest cities in the region at the upper tail of the city-size distribution (table 8.4).

DISCUSSION

The results of this analysis demonstrate that the hierarchical structure of urban systems is conservative, as theorized by Bessey (2002). This conclusion is true despite significant variability in the growth dynamics of individual cities (Bessey 2002). Membership of a city in a particular aggregation of cities may change over time, but this change does not alter the overall pattern of aggregations in regional city-size distributions. Further, changes in aggregation membership do not result in the "smoothing" of distributions. Rather, cities that move between aggregations do so in a discontinuous manner consistent with the conservative pattern of aggregations in the city-size distributions. For example, in 1890, Dallas was the largest city in the region, with Galveston and Houston as two of its rival cities within the same aggregation. Phoenix, in contrast, was a small town in 1890, with no indication of its meteoric rise during the next century. Over the course of the next few decades, Houston cemented its position of dominance in the region, Galveston began a slow descent to become a medium-size city, and Phoenix ascended to the third-largest city in the region by 1990. These cities demonstrate that change drives urban systems on a city level, but the underlying structure of aggregations remains robust to changes in these aggregations' membership.

Temporally discrete urban-growth rates (Papageorgiou 1980) and clumping in the spatial ranges of city functions (Korcelli 1977) provide clues to how spatially large systems—that is, national economies—"entrain" (Holling 1992) spatially smaller units, including regional and city economies, to produce stability in macrostructure but great diversity in the available growth paths (Dendrinos and Sonis 1990). These phenomena accord with the hierarchical structure that contributes to resilience, but contradict the neoclassical view regarding the nature of urban economics. The generally accepted stability of urban structure is interpreted as the manifestation of a steady-state condition in which a city-size distribution is affected by a myriad of small, random forces (Gabaix 1999). This conclusion assumes homogeneity in underlying growth processes; that is, growth is independent of city size, which appears inconsistent with the empirical data.

Similarly, in terms of the city-scale urban-ecology interface, Makse, Havlin, and Stanley (1995) have shown that few ecologically driven policies aimed at urban-exurban "greenbelt" preservation have met with success, perhaps because at some scale a city's growth dynamic operates outside the influence of exogenous "top-down" controls. Evidence from models points to incremental changes in some critical variable such as export-based manufacturing output as a trigger to

TABLE 8.4 Cities in the Aggregation at the Upper Tail of the Southwestern Region Distribution by Decade

1860	1870	1880	1890	1900	1910	1920
New Orleans	New Orleans	New Orleans	New Orleans	New Orleans	New Orleans	New Orleans
Louisville, Ky.	Louisville, Ky.	Louisville, Ky.	Louisville, Ky.	Louisville, Ky.	Louisville, Ky.	Louisville, Ky.
Richmond, Va.	Richmond, Va.	Richmond, Va.	Richmond, Va.	Memphis, Tenn.	Atlanta, Ga.	Atlanta, Ga.
Charleston, S.C.	Charleston, S.C.		Nashville, Tenn.	Atlanta, Ga.	Memphis, Tenn.	Birmingham, Ala.
	Memphis, Tenn.		Atlanta, Ga.	Richmond, Va.	Birmingham, Ala.	Norfolk, Va.
			Memphis, Tenn.	Nashville, Tenn.		Richmond, Va.
				Birmingham, Ala.		Memphis, Tenn.
				Norfolk, Va.		Nashville, Tenn.
				Wheeling, W.Va.		
				Savannah, Ga.		
				Charleston, S.C.		

1930	1940	1950	1960	1970	1980	1990
New Orleans	New Orleans	New Orleans	Miami-Hialeah, Fla.	Miami-Hialeah, Fla.	Atlanta, Ga.	Atlanta, Ga.
Louisville, Ky.	Atlanta, Ga.	Atlanta, Ga.	New Orleans	Atlanta, Ga.	Miami-Hialeah, Fla.	Miami-Hialeah, Fla.
Atlanta, Ga.	Louisville, Ky.	Louisville, Ky.	Atlanta, Ga.	New Orleans	New Orleans	Tampa, Fla.
Memphis, Tenn.	Birmingham, Ala.	Miami-Hialeah, Fla.	Louisville, Ky.	Louisville, Ky.	Fort Lauderdale, Fla.	Norfolk, Va.
Birmingham, Ala.	Memphis, Tenn.	Birmingham, Ala.	Memphis, Tenn.	Norfolk, Va.	St. Petersburg, Fla.	Fort Lauderdale, Fla.
		Memphis, Tenn.	Birmingham, Ala.	Memphis, Tenn.	Memphis, Tenn.	New Orleans
		Norfolk, Va.	Norfolk, Va.	Fort Lauderdale, Fla.	Norfolk, Va.	Orlando, Fla.
				Birmingham, Ala.	Louisville, Ky.	Memphis, Tenn.
				Jacksonville, Fla.		West Palm Beach, Fla.
				St. Petersburg, Fla.		Louisville, Ky.
				Nashville, Tenn.		Jacksonville, Fla.
				Richmond, Va.		Birmingham, Ala.
				Tampa, Fla.		Richmond, Va.
						Nashville, Tenn.

spontaneous population growth in cities once some threshold level is reached. When a region's transition to a high-level of urbanization occurs more rapidly than local or regional ecosystems' ability to absorb and sustain such growth, an understanding of the self-organizing properties of urban systems across scales becomes critical to management of tightly coupled ecological and human systems.

Holling (1992) and Dendrinos (1992; see also Dendrinos and Sonis 1990) have proposed that discontinuously distributed variables of inherently different speeds entrain growth processes to create clustering of species or cities in aggregations based on the similarity of their sizes. The evidence comes in the form of constraints on the size-density function, revealed as aggregations of species or cities, separated by gaps, which cannot be satisfactorily explained by developmental, historical, or other factors (Allen and Holling 2002; Bessey 2002). These discontinuities represent a departure from scaling laws typically represented by a simple continuous, unimodal distribution (Allen, Forys, and Holling 1999; Bessey 2002). Krugman (1996), too, has questioned Zipf's law as an explanation for growth processes in urban systems, arguing that Zipf's law is scale independent, which in turn suggests that hierarchical relationships of variables operating at different speeds play no part in shaping city sizes (see also Bessey 2002). City size may also be reflected in the services offered by a particular city. Is this conclusion accurate? Ostrom (1972) has suggested that neither economies nor diseconomies of scale are significant for cities of 25,000 to 250,000 individuals. However, in cities with a population greater than 250,000, size is a factor because there are significant diseconomies of scale.

Bessey (2002) has theorized that the spacing of cities on a national scale is driven by a slow dynamic. Cities may appear and disappear through time in response to locational advantages in a landscape. Human-ecological systems (e.g., cities) self-organize, and the manifestation of size (e.g., population) reflects the limitations of the landscape (Berkes and Folke 1998). For example, the rise of a city such as Phoenix may have been the result of a vacuum of urbanization in the southwestern region of the United States, combined with access to a critical resource (e.g., water) for city growth and development. On a smaller, regional scale, a fast variable driven by the minimum population and income needed for city survival also influences city size (Bessey 2002). This hierarchy of processes operating at different scales may allow urban systems to remain resilient in the face of disturbance (e.g., loss of a major industry) (Holling 1973; Bessey 2002). This hierarchical, multiscale relationship of processes indicates that there are fundamentally nonlinear interdependencies between growth rates and size (Bessey 2002). This structure is constant despite differences in city-size growth, but also despite the different membership of aggregations and members' respective positions within aggregations.

It is apparent that despite differing developmental histories, regional urban systems in the southwestern and southeastern United States concentrate population in the regions' largest cities. Jefferson (1939) stated that a country or a region's leading city is disproportionately large and expressive of the national or regional capacity and feeling. This "primate city" was thought to be at least twice as large as the next-largest city and more than twice as significant in terms of sociopolitical and economic importance. Although "primate cities" do not exist in the southwestern or southeastern regions of the United States, each region has two or more dominant cities within aggregations at the upper tier of city-size distributions. Similar to Bessey's (2002) results, there was increasing population concentration in large cities in the southwestern and southeastern regions of the United States over one hundred years. This pattern manifested in a discontinuous structure in the city-size distributions with pronounced gaps between aggregations at the upper tail of decadal data sets.

This study demonstrates that there are departures from scaling laws in city-size distributions for the southwestern and southeastern regions of the United States. These patterns of aggregation in city-size distributions were constant throughout the twentieth century, despite consistent change in the membership of individual aggregations (figs. 8.26 and 8.27). Our analysis indicates that an important pattern in urban system distributions has been ignored in the desire to fit city-size distributions to power laws and that these systems' structures and dynamics appear to be more dynamic than recent research has indicated.

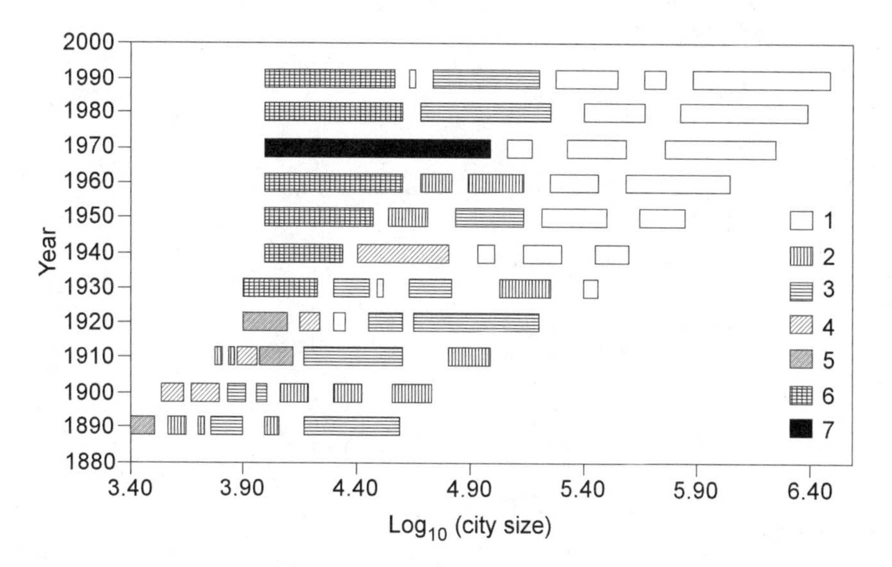

FIGURE 8.26 Location of discontinuities in the city-size distributions for the southwestern region of the United States from 1890 to 1990. The different shades indicate the percentage of cities within an aggregation: 1 (0–5%), 2 (5–10%), 3 (10–20%), 4 (20–40%), 5 (40–60%), 6 (60–80%), and 7 (80–100%).

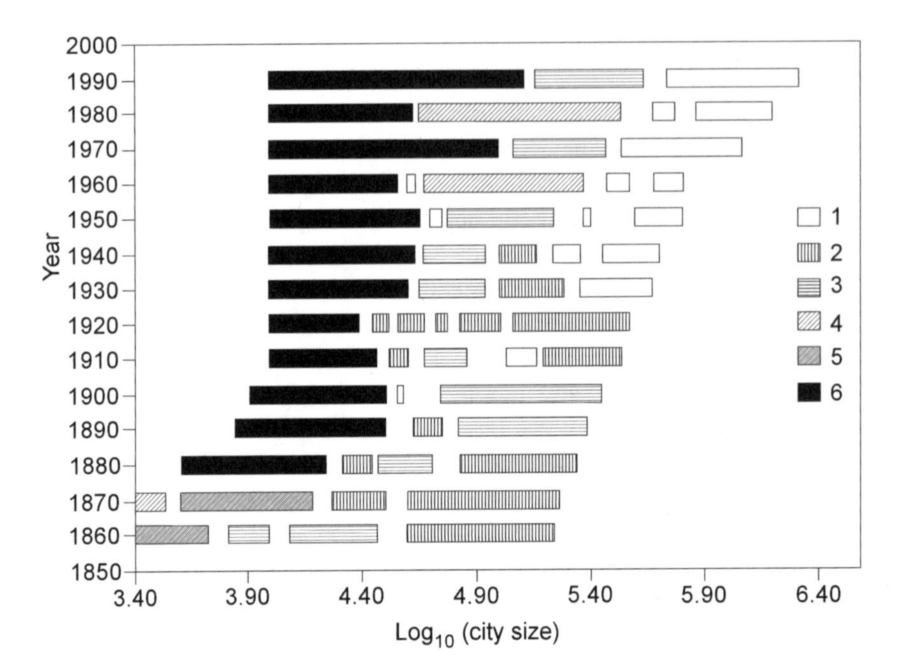

FIGURE 8.27 Location of discontinuities in the city-size distributions for the southeastern region of the United States from 1860 to 1990. The different shades indicate the percentage of cities within an aggregation: 1 (0–5%), 2 (5–10%), 3 (10–20%), 4 (20–40%), 5 (40–60%), and 6 (60–100%).

9

EVALUATING THE TEXTURAL DISCONTINUITY HYPOTHESIS

A Case for Adaptive Inference

Craig A. Stow, Jan P. Sendzimir, and Crawford S. Holling

TEMPORAL AND spatial patterns in landscapes are controlled by key structuring processes, each functioning at characteristic periodicity and spatial scales (Holling 1992). This hierarchical structure (T. Allen and Starr 1982; O'Neill et al. 1986) produces discontinuous patterns on the landscape and a clumpy or uneven distribution of its resources across the scales over which animals live. This uneven, discontinuous resource distribution provides variation in the type and amount of ecological opportunity available to species of different body sizes. Small animals depend on the patterns and resources over small-scale ranges; large animals over large-scale ranges. As a consequence, Holling (1992) proposed that animals have evolved physical and behavioral characteristics to exploit an environment with a texture that varies across scales and that species will assort themselves and adapt to the specific patterns of any specific landscape. Holling's (1992) Textural Discontinuity Hypothesis (TDH) posits that animal body masses are entrained to the specific pattern and structure available in a given system at a given scale. Because that landscape structure is discontinuous in the distribution of its resources across scales, then animal body-mass distributions should reflect this discontinuity by exhibiting a clump-and-gap (discontinuous) pattern. This idea has potentially profound ecological and management implications, including the recent discovery that both endangered and invasive species tend to fall near the edges of the species clusters of similar body masses (Allen, Forys, and Holling 1999; Allen 2006; chap. 11 in this volume).

However, in addition to stirring considerable interest, the proposal has evoked some skepticism. Brown (1995) agrees that body-mass distributions appear discontinuous, but offers an alternative explanation. Manly (1996) asserts that body-size distributions are not multimodal, but rather are plausible small-sample outcomes from continuous unimodal or, at most, bimodal distributions.

We thank Dave Otis and Steve Carpenter for reviewing earlier versions of this manuscript.

Siemann and Brown (1999) argue against the existence of clumps in mammal data sets by analyzing the gap size between clumps, concluding that the observed gaps are not unusually large when compared to gaps generated from a continuous uniform distribution.

The desire to rigorously explore the TDH's premise and implications has led us to deliberate the nature of the data on which the hypothesis is based and to explore a variety of statistical methods and interpretations. We have considered the statistical approaches offered in the published challenges and have evaluated methods that we believe more accurately capture the TDH's premise. In this analysis, we evaluate the arguments and approaches offered by Manly (1996) and by Siemann and Brown (1999) and then propose an alternative. We suggest that adaptive inference (Holling and Allen 2002) provides a better framework than a narrow classical hypothesis testing structure for investigating the TDH's implications.

EXAMINING THE NUMBER OF MODES

The TDH arose initially from the visual observations that adult animal body-mass histograms appear discontinuous and jagged rather than smooth and that this pattern occurs consistently in different systems (Holling 1992). Manly (1996) examined this supposition as a question of multimodality, asking: "Do the body-mass distributions provide evidence for more than one mode?" He posed this question of the data sets originally examined by Holling, using Silverman's (1981) kernel density method, and concluded that there was little evidence for more than two modes in any of the data examined.

Determining the location of modes in a distribution consisting of several component distributions is a difficult problem, particularly when the number of modes is unknown. Extensive statistical literature on the subject explores the properties of various methodologies. The problem becomes increasingly difficult as the number of suspected modes increases and the distance between modes decreases (Roeder 1994). Silverman's method is known to estimate conservatively the actual number of modes. In other words, in the test of the null hypothesis, that the data likely arise from a distribution with one mode, Silverman's test will overestimate *P*-values, even in large samples (Hall and York 2001). Hall and York show in the simplest case, with an H_A of two modes, the power of test is low until modes become widely separated. Thus, the number of modes is likely to be underestimated using this method.

In addition, Manly (1996) clearly thinks that his inquiry draws inferences about the underlying distributions of the body-mass data, implying that the available data are small samples from a large population. However, we argue that the data we have examined so far are not appropriately viewed as small samples; rather, they represent censuses, or nearly so, of specific geographic regions. We have considered relatively large, easily observable species for which a significant amount of undercounting is unlikely. Although there may be some measurement error in the estimation of the mean adult body mass for each species, this error is small relative to the total variance of the body-mass distribution. Measurement error also tends to obscure pattern (fig. 9.1). A well-known example is the least-squares slope estimator's bias toward zero (the obfuscation of a relationship between response and predictor variable) in a simple linear regression, when the predictor variable is observed with error (Fuller 1987). Therefore, we do not consider statistical testing of these data to be telling us whether or not the observed discontinuities and aggregations are "real" or just artifacts of sampling error. We believe the data to be a close representation of the full body-mass assemblages and that it is more appropriate and interesting to ask, "What causes these patterns?" rather than "What are the real but unseen patterns?" As this inquiry is extended to body-mass assemblages where the near-census assumption is clearly not correct, then drawing inferences to locate probable modes becomes more appropriate (e.g., Havlicek and Carpenter 2001).

EXAMINING GAP SIZE

Siemann and Brown (1999) approached this question differently than Manly, examining the size of the gaps (discontinuities) within body-mass distributions rather than the number of modes. They simulated adult body sizes using pseudorandom values from uniform distributions. They tailored uniform distributions to conform to animal data sets from several biomes. In repeated sampling, they showed that the biggest gap observed in most of the data sets examined had a high *P*-value (it was not in the tails of the simulated distribution), indicating that it was a plausible outcome from a continuous uniform process. In other words, a high *P*-value for the biggest real gap was considered evidence that the gaps in the real data sets were not surprisingly large. In the few instances where the biggest gap did have a small *P*-value, they successively compared the second- and third-biggest real gaps to the simulated distribution of gaps to show that there were not unexpectedly large multiple gaps. From this analysis, Siemann and Brown concluded that gappiness was not a general feature of mammalian data sets.

This testing is weak, both conceptually and methodologically. From a conceptual perspective, it is not clear what the comparison of a small random sample generated by a uniform process is actually testing. We have argued that our data sets characterize more than just small samples from much larger, unseen

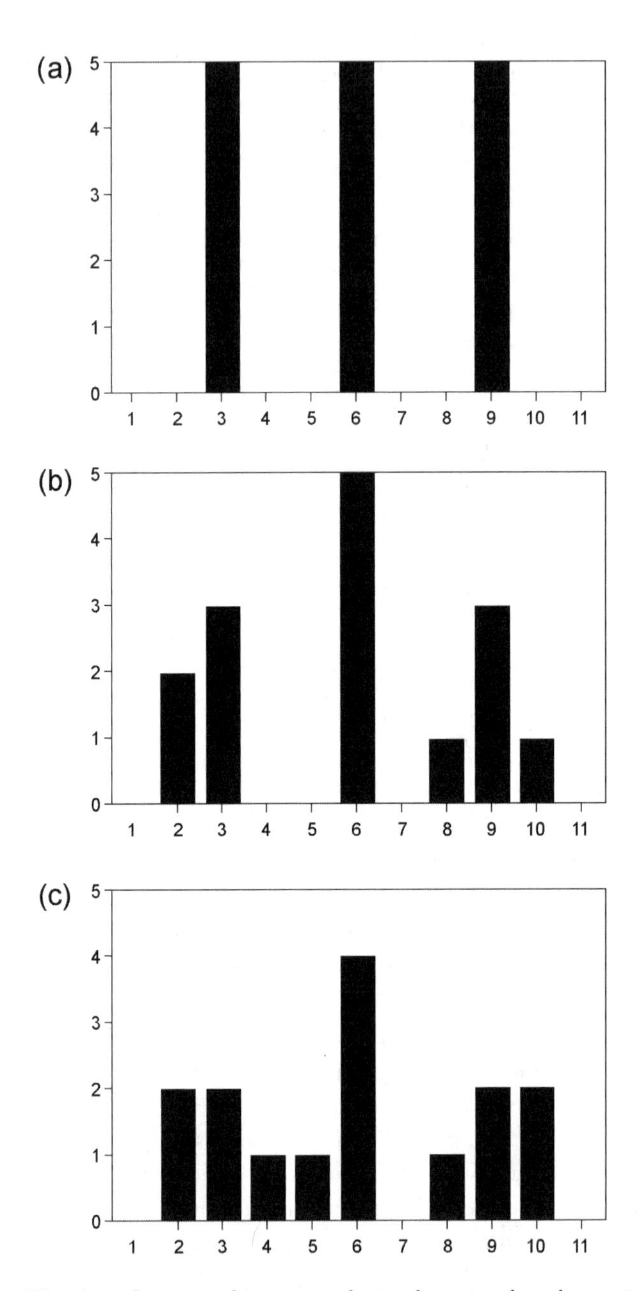

FIGURE 9.1 The three frequency histograms depict the point that observation error of increasingly high variance tends to obscure the pattern in the data. The *top panel (a)* shows idealized original data with no observation error clearly defining the three aggregations. The *middle panel (b)* is the same data with normal mean zero, variance 0.1 data added to each observation. The *bottom panel (c)* depicts the same data with normal mean zero, variance 0.5 data added to each observation.

populations and contend that the data sets are censuses or near censuses in most cases. Indeed, could the TDH have originated from empirical observation without this implicit assumption? Small random samples from a uniform distribution are not evenly distributed; they usually exhibit gaps and clumps (fig. 9.2). Concluding that body-mass distributions are not different from these small-sample realizations from a uniform null is not a strong test of the TDH, particularly without also conducting analogous tests using multimodal distributions.

Our biggest criticism, however, is that the procedure Siemann and Brown used is methodologically flawed. Considering first the biggest gap and then a conditional sequence of successively smaller gaps does not accurately capture the TDH's premise. Holling's (1992) thesis is based on the idea that adult body-size distributions exhibit an *assemblage* of discontinuities and aggregations that provide clues regarding underlying scaling processes. The hypothesis is not premised on the idea that the biggest gap is unusually large. Siemann and Brown's method stops if the biggest gap from a particular species list is not unusually large, with the conclusion that those species are a plausible realization from a continuous uniform distribution. However, even if the biggest gap is not unusually large, the complete set of gaps may not represent a likely outcome from a continuous uniform process. Therefore, it is pertinent to consider the assemblage of gaps simultaneously rather than to consider gaps sequentially from largest to smallest.

In addition to not accurately capturing the TDH's premise, the sequential nature of Siemann and Brown's testing almost guarantees that few species lists

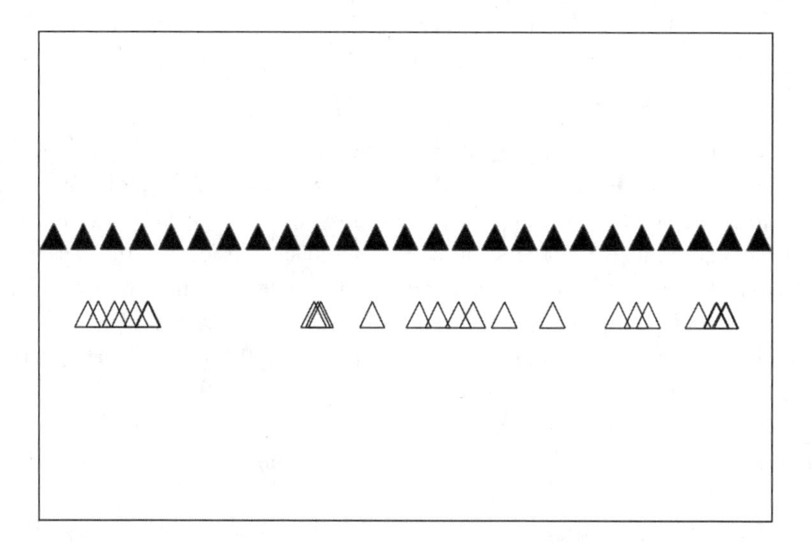

FIGURE 9.2 Comparison of samples ($n = 25$) randomly generated from a uniform distribution (△) to a systematically generated, evenly spaced sample (▲).

will be found to have a set of unusually large gaps. These authors tested the second-biggest gap only if the biggest gap was judged to be unusually large. Then, if the second-biggest gap also was judged to be unusually large, they tested the third-biggest gap, and so on. The problem arises because the sum of all the gaps is constrained to equal the interval width between the least and greatest body masses. Thus, if the biggest gap is unusually large, it becomes less likely that the second-biggest gap will also be unusually large because the it can be no larger than the difference between the interval width and the largest gap. If both the biggest and second-biggest gaps are unusually large, then it is increasingly less likely that subsequent gaps will be unusually large. In the most extreme example, when all body masses lie on one end or the other of the distribution, then the biggest gap will equal the entire interval width, and all other gaps will be of size zero, resulting in only one "significant" gap. Thus, the sequential testing performed by Siemann and Brown is unlikely to support the presence of an assemblage of large gaps. In essence, they tested a sequence of conditional probabilities that are guaranteed to approach zero.

To make the approach initiated by Siemann and Brown more consistent with the TDH's premise, we evaluated whether or not the entire set of gaps appeared to be a likely outcome from a continuous uniform distribution, which can be evaluated by examining the distribution of the length, or norm (Friedberg and Insel 1986), of the entire vector of gaps. The norm is calculated as:

$$\sqrt{\sum_{i=1}^{m} \text{gap}_i^2},$$

where m is the number of gaps ($n-1$, where n is the number of species) in the geographic region being evaluated. It is appropriate to square the individual gaps before summing them because simply summing them would always generate a value equal to the total interval width. This approach considers all the gaps from a given geographic region simultaneously rather than sequentially. For a given interval, the largest possible value of our test statistic occurs if all species lie on either end of the interval. In this case, there is one gap that constitutes the entire interval width, and the rest of the gaps are zero. The smallest possible value of our test statistic occurs if the species are evenly distributed and all of the gaps are the same size. Therefore, if the body-mass distributions are indeed discontinuous, this statistic will tend to be large.

We did this analysis for the six species lists in the appendix of Holling's 1992 paper and for the mammalian species lists from twenty-one North

American biomes presented by Brown and Nicoletto (1991). For each species list, we drew two thousand simulated species lists from a continuous uniform distribution, with endpoints conforming to the log of the smallest and largest body masses from the corresponding real list. Each simulated species list consisted of n species, where n was the number of species in the real list. Because Siemann and Brown used $n-2$ species instead of n in their simulations, we also repeated each analysis using $n-2$. *P*-values were calculated as the proportion of each set of two thousand draws where the simulated gap norm exceeded the real gap norm.

Calculated *P*-values were slightly greater when $n-2$ species rather than n species were generated for each simulated species list, but the results are qualitatively similar (table 9.1). Because we believe that n is the more appropriate value, we focus our discussion on these results. The gap norm in twelve of the twenty-one mammalian data sets from Brown and Nicoletto (1991) had a *P*-value of less than 0.05, consistent with classical "significance" (fig. 9.2). Eighteen of the twenty-one data sets had *P*-values of less than 0.3, but only three had *P*-values greater than 0.5, falling in the lower half of the distribution. None of the three mammal data sets from Holling 1992 was significant by this measure, but all were in the upper half of the distribution. Two of the three bird

TABLE 9.1 Animal Data Sets Used to Test for Discontinuities Using the Vector Norm

GEOGRAPHIC REGION	*P*-VALUE, *N*	*P*-VALUE, *N*-2
North America (all)	0.000	0.000
Sitkan	0.275	0.463
Oregonian	0.010	0.021
Yukon taiga	0.676	0.899
Canadian taiga	0.239	0.373
Eastern forest	0.236	0.368
Austroriparian	0.172	0.306
Californian	0.011	0.025
Sonoran	0.001	0.001
Chihuahuan	0.001	0.003

TABLE 9.1 *Continued*

GEOGRAPHIC REGION	*P*-VALUE, *N*	*P*-VALUE, *N* - 2
Tamaulipan	0.082	0.153
Great Basin	0.010	0.019
Alaskan tundra	0.650	0.671
Canadian tundra	0.538	0.589
Grasslands	0.003	0.012
Rocky Mountains	0.041	0.074
Sierra-Cascade	0.002	0.003
Madrean-Cordilleran	0.000	0.000
Campechean	0.000	0.000
Guerreran	0.000	0.000
Sinaloan	0.068	0.119
Yucatecan	0.023	0.052
Boreal Forest Mammals	0.188	0.379
Boreal Prairie Mammals	0.167	0.330
Boreal Forest Birds	0.000	0.000
Boreal Prairie Birds	0.157	0.257
Pelagic Northwestern Birds	0.001	0.001
Antelopes	0.058	0.137

Sources: The first twenty-two entries are mammal communities taken from Brown and Nicoletto 1991; the last six entries are mammal and bird communities from Holling 1992.

assemblages from Holling 1992 were significant, but the third was in the upper tail of the distribution.

Overall, twelve of the twenty-four mammalian data sets exhibited "significant" *P*-values, whereas twenty-one of the twenty-four were in the upper half of the distribution (an aggregate binomial *P*-value of 0.00014). Including the bird assemblages, fourteen of the twenty-seven data sets were "significant," and twenty-four of the twenty-seven were in the upper half of the distribution (an aggregate binomial *P*-value of 0.000025).

In contrast to Siemann and Brown, who concluded that there is "no general pattern of multiple gaps and clumps of body sizes in terrestrial mammal assemblages at the scale of biomes" (1999, 2791), our analysis of the same data indicates that a discontinuous pattern is common. We believe that the test we have used captures more appropriately the TDH's substance than the test used by Siemann and Brown. In two of the three bird assemblages explored, boreal forest and pelagic North American birds, *P*-values for the largest gaps are well below standard thresholds of "significance," and it has been suggested elsewhere that bird body-mass distributions are nonrandom (Maurer 1998a, 1998b).

A feature worth noting in these results is that the calculated *P*-values exhibit some relationship to the number of species for that particular geographic region, with regions of larger *n* having lower *P*-values (figs. 9.3 and 9.4). This relationship suggests that there may be a power effect, making it difficult to discern deviations from continuous uniformity in geographic regions with low species counts. In particular, all three geographic regions that exhibited *P*-values of greater than 0.5 had a low *n*. Indeed, other researchers have noted this result and utilized methods that maintain constant power rather than constant alpha when comparing body-mass distributions with widely varying numbers of species (Allen, Forys, and Holling 1999).

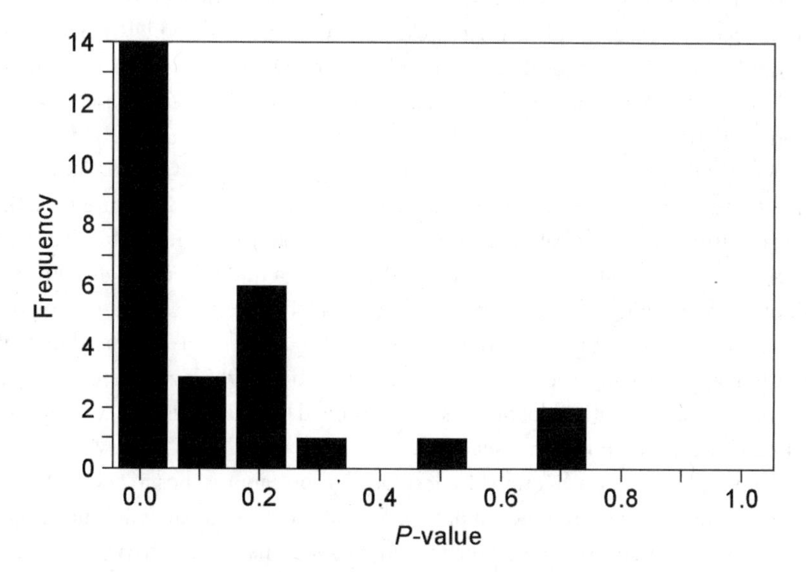

FIGURE 9.3 Histogram of the calculated *P*-values from all twenty-seven data sets used to test for patterns of discontinuity by means of the vector norm.

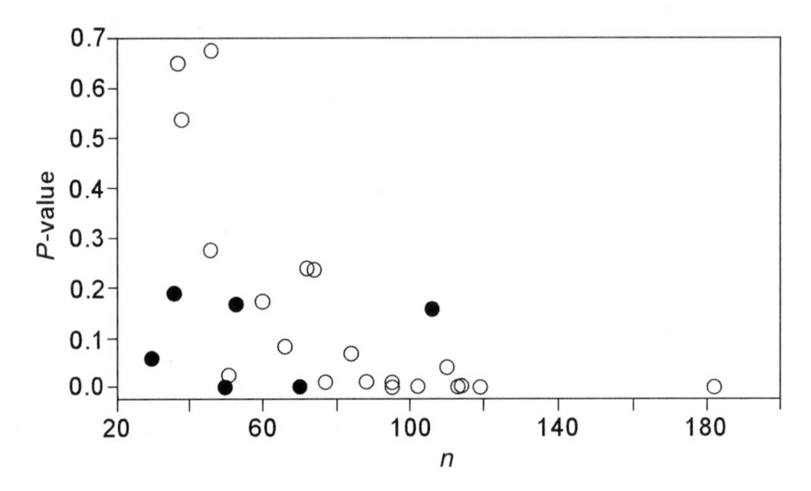

FIGURE 9.4 Bivariate plot of *P*-value versus the number of species. (○) Data from Brown and Nicolleto 1991. (●) Data from Holling 1992.

We emphasize that the use of a continuous uniform null model is primarily a test for nonuniformity. We have extended this model's application in this analysis because it was the null model used by Siemann and Brown. Many other null models are conceivable. It is possible to choose a particular null based on properties of the observed data, such as the mean and variance, but subsequently testing these same data against the null distribution tailored to mimic them becomes a conundrum. Employing a uniform null model is not an attempt to enumerate or locate gaps and clumps in a given data set; it is intended only to help differentiate between nonrandomness and a visual perception of nonrandomness.

The TDH presents a rich fabric of multiple hypotheses to evaluate. Squeezing this process into an overly narrow classical hypothesis-testing structure runs the risk of turning the exploration into a succession of potshots at nondefinitive straw men. Discerning pattern from randomness can be a question of resolution; random number generators, such as the one used in this analysis, are completely deterministic, but if their underlying processes are not known, they produce output that appears patternless. Similarly, exploring the TDH invites researchers to probe whether identifiable processes underlie data that may appear random when viewed at inappropriate scales.

However, there is still a need to develop rigorous approaches to assess the apparent structure in these kinds of data sets and to make comparisons among data sets so that the interpretation of results is not capricious. Exploration of the use of methods including cluster analysis as well as classification and regression trees

(Breiman et al. 1984) and their Bayesian implementation (Chipman, George, and McCulloch 1998) is ongoing and promising. We suggest that the application of such procedures in an adaptive inference framework (Holling and Allen 2002) is most likely to facilitate methodological progress and understanding of the mechanisms responsible for the observed dynamics and structures in complex systems.

PART THREE

CONSEQUENCES

10

DYNAMIC DISCONTINUITIES IN ECOLOGIC-ECONOMIC SYSTEMS

J. Barkley Rosser Jr.

A Public Domain, once a velvet carpet of rich buffalo-grass and grama, now an illimitable waste of rattlesnake-bush and tumbleweed, too impoverished to be accepted as a gift by the states within which it lies. Why? Because the ecology of the Southwest happened to be set on a hair trigger.

—A. LEOPOLD, "THE CONSERVATION ETHIC" (1933)

WITHOUT DOUBT prior to the appearance of human beings, ecological systems experienced discontinuities in their dynamic paths, if only due to dramatic exogenous shocks such as meteorite strikes or volcanic eruptions that probably triggered the episodes of mass extinctions known to have occurred in geological time. We also now know that in contrast to Darwin's view that "natura non facit saltum" (nature does not take a leap) (1895, 166), evolution may well have proceeded in a more discontinuous, "saltationalist," manner via punctuated equilibria, with long periods of little change alternating with periods of very rapid change (Gould 2002). However, it is also likely that these periods of rapid evolutionary change were driven by sudden changes in broader environmental conditions such as climate.

Of somewhat greater interest are the fluctuations and discontinuities that arise in an endogenous manner from within ecosystems. The Lotka-Volterra predator-prey cycle is perhaps the most famous and widespread of such phenomena, although it does not involve discontinuities per se (Lotka 1925). Two (or more) species interact in a coupled oscillation whose swings can be so great that they can appear to be almost discontinuous at certain points, as in the famous Hudson Bay hare-lynx example (Odum 1953; but for a dissenting view that such cycles are really driven by maternal effect dynamics in the prey population, see Ginzburg and Colyvan 2004). An even more dramatic example is the spruce budworm dynamics in Canadian forests (Holling 1965; May 1977; Ludwig, Jones, and Holling 1978; Ludwig, Walker, and Holling 2002), the roughly forty-year cycle marked by discontinuous outbreaks of the budworm population. The latter clearly involves a situation of multiple equilibria with possible hysteresis cycles between them.

Other strictly ecological systems that exhibit such multiple equilibria with accompanying discontinuities as systems move from one basin of attraction to another include coral reefs (Done 1992; Hughes 1994), kelp forests (Estes and Duggins 1995) and potentially eutrophic shallow lakes (Schindler 1990; Scheffer 1998; Carpenter, Ludwig, and Brock 1999; Carpenter, Brock, and Ludwig 2002; Wagener 2003), although the latter can be affected by human input of phosphorus. A more particular phenomenon involves situations when species invade ecosystems, leading to sudden disruption, as in some grasslands (D'Antonio and Vitousek 1992), although these invasions have increasingly been triggered by humans' transporting species across Wallace's Realms (Elton 1958).

The focus in this chapter is on systems in which human activities are crucial in the dynamics, especially through actions driven by economic motives. Hence, I consider dynamic discontinuities of combined ecologic-economic systems, in particular fishery collapses and discontinuities in forestry-management systems. However, the problem of lake eutrophication due to phosphorus loadings from human activity noted earlier is another important example, as is the problem of overgrazing of ecologically fragile rangelands that can be taken over by woody vegetation (Noy-Meir 1973; Walker et al. 1981; Ludwig, Walker, and Holling 2002). A fundamental concept in this discussion is the stability-resilience trade-off first posited by Holling (1973), with true discontinuities being associated with crossing from one basin of attraction to another in a failure of resilience.

FISHERY COLLAPSES

In *Bioeconomic Modelling and Fisheries Management*, Colin W. Clark lists species that have experienced relatively sudden population collapse in fisheries: Antarctic blue whales, Antarctic fin whales, Hokkaido herring, Peruvian anchoveta, Southwest African pilchard, North Sea herring, California sardine, Georges Bank herring, and Japanese sardine (1985, 6). Although some of these species have since recovered, others have more recently collapsed, most dramatically the Georges Bank cod. Needless to say, these events have drawn much study and effort, with many approaches being used, including catastrophe theory for the Antarctic blue and fin whale cases (Jones and Walters 1976; for a more thorough discussion of the application of catastrophe theory in economic models, see Rosser 1991). A more general approach has been developed (C. Clark 1990; Hommes and Rosser 2001; Rosser 2001) based on the study of open-access fisheries by Gordon (1954), combined with Shaefer's (1957) yield curve model and drawing on Copes's (1970) insight that the supply curve in many fisheries may be backward bending. However, this more general approach also applies to closed-access fisheries managed in an optimal manner, the open- and closed-access cases coinciding for the case where the discount rate for the future is infinite—that is, where the future is not counted at all.

Let x = fish biomass, r = intrinsic growth rate of the fish, k = ecological carrying capacity, t = time, h = harvest, and $F(x) = dx/dt$, the growth rate of the fish without harvest. Then a sustained yield harvest is given by

$$h = F(x) = rx(1 - x/k). \qquad (1)$$

Let E = catch effort in standardized vessel time, q = catchability per vessel per day, c = a constant marginal cost, p = price of fish, and δ = the time discount rate. The basic harvest yield is

$$h(x) = qEx. \qquad (2)$$

Hommes and Rosser (2001) show that the optimal discounted supply curve is then given by

$$x\delta(p) = k/4\{1 + (c/pqk) - (\delta/r) + [(1 + (c/pqk) - (\delta/r)^2 + (8c\delta/pqkr)]^{1/2}\}. \qquad (3)$$

This entire system is depicted in figure 10.1, with the backward-bending supply curve in the upper right and the yield curve in the lower right. It must be noted that if the discount rate is low enough, then there is no backward bend, although the bend will happen for discount rates exceeding 2%, which is not very high. The maximum backward bend will occur with a discount rate of infinity, in which the future is not counted at all, which coincides with the open-access case.

The basic collapse story occurs as a sufficiently inelastic demand curve gradually shifts outward. This is also shown in figure 10.1, and it can be seen that as the demand curve shifts outward, the system will go from a single equilibrium with a low price and large fish stock and harvest to a zone of three equilibria and then will suddenly jump to another single equilibrium case of high price and low fish stock and harvest as the demand curve continues to move outwards (Anderson 1973). Needless to say, this scenario is more likely for the case of open access with its known propensity for overexploitation of a natural resource. Hommes and Rosser (2001) studied the emergence of chaotic dynamics in the zone of multiple equilibria with naive adjustment by boundedly rational fishers. In effect, the fisheries bounce back and forth between the two competing stable (outer) equilibria.

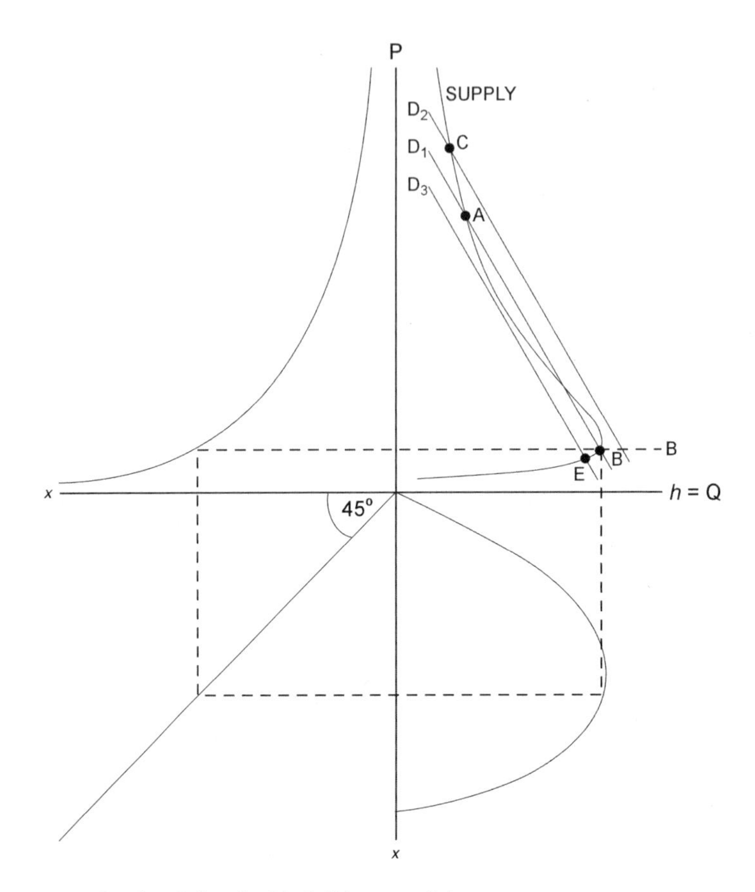

FIGURE 10.1 Gordon-Schaefer-Clark fishery model.

It is possible for them to mimic an underlying chaotic dynamic with a simple autoregressive adjustment rule. Conklin and Kolberg (1994) have also studied chaotic dynamics in fisheries.

The scenario given here does not generate a complete collapse of the fishery to zero. Such a collapse can happen if the yield function exhibits the character of depensation (Allee 1931; C. Clark and Mangel 1979). The yield function and the harvest yield curve for the critical depensation case are shown in figure 10.2. In this case, a complete collapse of the population to zero can happen—the situation when a species has a minimum number below which it will fail to reproduce.

The potential for complete collapse of a fishery can also arise from the phenomenon of capital stock inertia (C. Clark, Clarke, and Munro 1979). This potential is exacerbated in practice by many fishers' unwillingness to give up their lifestyle and the tendency for many to belong to isolated minority groups with strong identities who resent policy efforts by outsiders (Charles 1988). It is well known that if fishers

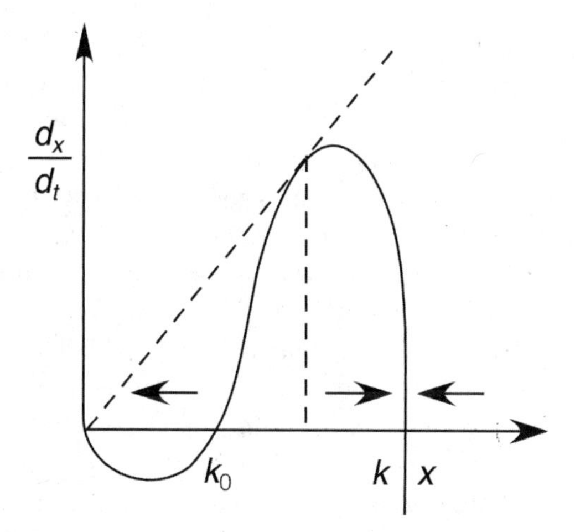

FIGURE 10.2 Critical depensation yield function.

can organize themselves to manage their fisheries and to control access, the outcome is superior to when outsiders attempt to manage the fisheries (Walters 1986; Durrenberger and Pàlsson 1987; Ostrom 1990; Bromley 1991; Sethi and Somanathan 1996). However, game theoretic approaches are relevant in the analysis of conflicts over access, especially in connection with fisheries spilling across national boundaries (Okuguchi and Svidarovsky 2000; Bischi and Kopel 2002). The problem is not "common property," as was thought in the period when Gordon (1954) wrote, but rather "control of access" (Ciriacy-Wantrup and Bishop 1975).

Yet another potential cause of the collapse of a fish species involves interactions between competitive species. Gause (1935) developed the competitive species version of the Lotka-Volterra model. Let s = intrinsic growth rate of the y species and L its carrying capacity, with all other variables defined as given earlier, so that

$$dx/dt = rx(1 - x/k) - \alpha xy. \qquad (4)$$

$$dy/dt = sy(1 - y/L) - \beta xy. \qquad (5)$$

An isocline for x to remain constant is given by

$$Y = (r/\alpha)(1 - x/k) - (q/\alpha)E. \qquad (6)$$

As E increases, the system can bifurcate, with a collapse of x occurring. This bifurcation will coincide with x's competitive replacement by y when $E = s/\beta$ (C. Clark 1990). Murphy (1967) analyzed the collapse of the Pacific sardine fishery in the late 1940s and its replacement by the anchoveta as a likely example of such a competitor-driven collapse. Figure 10.3 depicts this case, with x collapsing while its yield-effort curve is still increasing.

In contrast to the competitor species case, there is the question of dynamics in the predator-prey case. Brauer and Soudack (1979) studied this case with regard to the harvesting of the predator species. They found a stable case and an unstable case, the latter emerging as the harvest rate increases. In both cases, there are two equilibria. In the stable case, one equilibrium is stable, but the other is an unstable node from which the dynamics move to the stable equilibrium. In the unstable case, the formerly stable equilibrium destabilizes, and the collapse of the predator population is possible. This situation is depicted in figure 10.4, with 10.4a the stable case and 10.4b the unstable case. Brauer and Soudack (1985) suggested various strategies for stabilizing the unstable case.

Carl Walters (1986) has also studied the problem of managing fisheries with such predator-prey interactions. More particularly, he has considered the problem of preserving trout in Lake Superior after their devastation by the lamprey, which invaded through the St. Lawrence Seaway. He has suggested that in an effort to maximize yield while avoiding catastrophic collapse, one may have to "surf" near the edge of the system as in *(a)* in figure 10.5. Fishers face the trade-off between too little information but a higher and safer yield of the wild trout population, on the one hand, and more information but the danger of a catastrophic collapse, on the other, as in *(b)* in figure 10.5.

FOREST DYNAMIC DISCONTINUITIES

I have noted the case of the spruce budworm cycle and the influence that distant human activities can have on it (Holling 1988). I now consider certain situations where more direct human management of forests can lead to discontinuities of

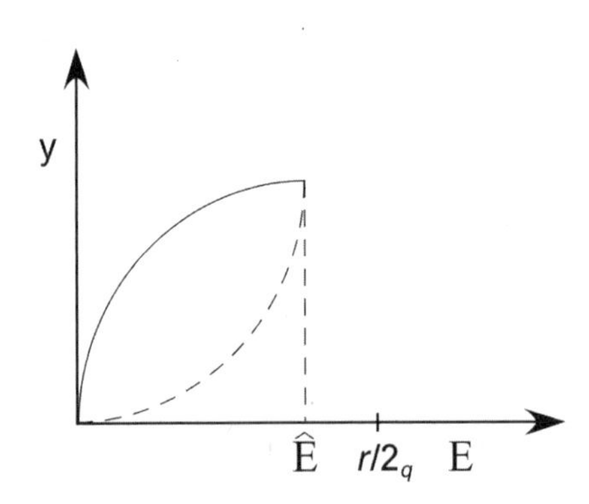

FIGURE 10.3 Fishery collapse with competitor species.

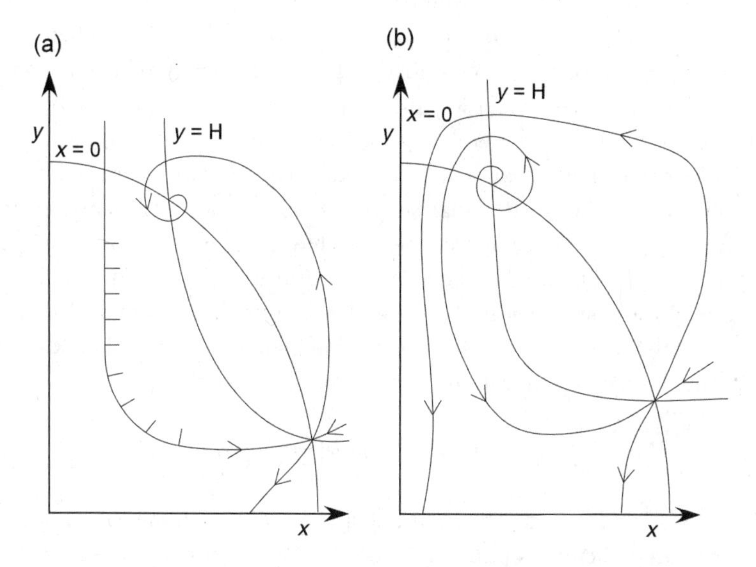

FIGURE 10.4 Complex predator-prey dynamics with harvesting: *(a)* stable situation; *(b)* unstable situation.

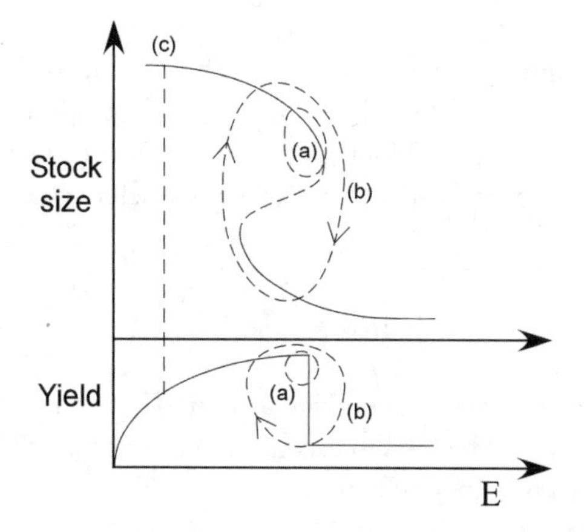

FIGURE 10.5 Great Lakes trout dynamics under alternative strategies.

various sorts. In this analysis, I assume that the forest managers are optimizing agents. Although in some cases this optimization may take the form of simple wealth maximization based on future expected income streams from timber harvesting or grazing of species with economic payoffs, it can be generalized to managers of publicly owned forests who are valuing benefits that do not have marketed economic value, such as various forms of recreation, carbon sequestration, biodiversity preservation, and so on.

Following Faustmann's ([1849] 1968) pioneering work that focused strictly on the optimal rotation period of a forest being utilized strictly for timber and replanted after harvesting, Hartman (1976) expanded the focus to include nontimber amenities. Let $f(t)$ = the growth function of the trees, p = the price of timber (assumed constant, but note that constancy of price is a nontrivial assumption; when price can change stochastically, options pricing theory can be used, which considerably complicates the analysis [Arrow and Fisher 1974; Zinkhan 1991; Saphores 2003]), r = the real market rate of interest, c = the marginal cost of timber harvesting, e = the base of the natural logarithms, $g(t)$ = the time pattern of the value of the nontimber amenities, and T = the optimal rotation period (the period that the forest is allowed to grow before being cut and replanted). Then the infinite horizon optimization gives the following solution:

$$pf'(T) = rpf(T) + r[(pf(T) - c)/(erT - 1)] - g(T). \qquad (7)$$

This solution implies that if the nontimber amenities continue to have positive value over time, then the optimal rotation will be longer than for a forest being managed solely for its timber value. Indeed, if $g(t)$ continues to grow over time and is sufficiently high, it can become optimal not to cut the trees, even though they have timber value.

However, if the $g(t)$ is high early in the life of the forest and declines, this scenario can lead to quite a different solution. Thus, many forests possess grazing benefits in early stages. Swallow, Parks, and Wear (1990) have studied this case for national forests in western Montana that possess cattle-grazing benefits in the early stages of forest growth. They estimated parameter values for the following grazing function:

$$g(t) = \beta_0 \exp(-\beta_1 t). \qquad (8)$$

The value of this function is maximized at $T = 1/\beta_1$, which for the case of cattle grazing in the western Montana forests was found to be at 12.5 years. The function they estimated is shown in figure 10.6.

Figure 10.7 depicts for this case the present value of the forest at different ages and the comparison between the marginal benefit of delaying harvest (MBD) and the marginal opportunity cost (MOC) of doing so. Clearly, there are multiple solutions, although in this case the later local optimum is superior. Nevertheless, for this case, the optimum occurs at an earlier time than the pure Faustmann solution would indicate in the absence of this early grazing benefit. Implicit here is the possibility of a discontinuity arising from a change in the market rate of interest, r, which influences both the MBD and the MOC. Thus, there might be drastic changes in harvest policy at a critical

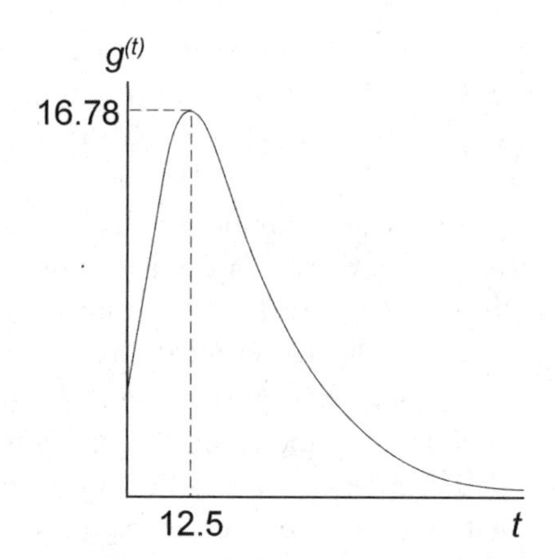

FIGURE 10.6 Grazing benefit function.

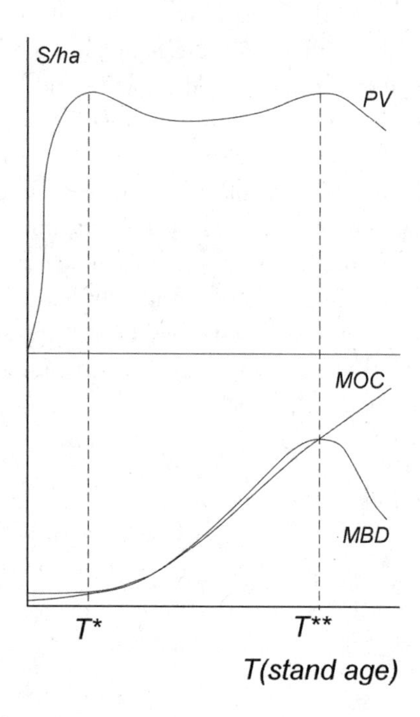

FIGURE 10.7 Optimal Hartman rotation (PV = present value; MBD = marginal benefit of delay; MOC = marginal opportunity cost).

interest rate, with the optimal rotation suddenly jumping to a very different value. This case is reminiscent of how discontinuities can arise when capital theoretic paradoxes such as reswitching occur in cases where time patterns of costs and benefits are irregular over time (Porter 1982; Rosser 1983; Prince and Rosser 1985).

A more nonmarket type of benefit is hunting. Figure 10.8 depicts the time pattern of $g(t)$ for hunting in the George Washington National Forest in Virginia as estimated by Rosser (2005) based on data gathered during a land-use planning exercise as part of the FORPLAN process in the early 1980s (Johnson, Jones, and Kent 1980). This pattern shows three separate peaks of hunting value at different times after a clear-cut of these largely deciduous forests. The first peak occurs about six years after the cut when deer hunting is maximized. This situation is somewhat analogous to the example of cattle grazing noted earlier, with the deer favoring the clear-cut zones with small trees and especially the edges of such zones. The next peak occurs around twenty-five years after the cut and is associated with hunting of wild turkeys and grouse. This period is approximately the forest's period of maximum biodiversity as the succession from oaks to successor species begins and there is much underbrush. Finally, after about sixty years, the forest becomes "old growth," and bear hunting is maximized, basically increasing in value with the age of the forest because the bears like large downed trees (the forest supervisor at that time made clear to me that the biggest conflict he had to deal with between different groups in the public was between the deer hunters and the bear hunters, both very well-organized and articulate groups, whose interests regarding how much timber harvesting should occur were, and are, very much at odds). In many less-developed countries, the hunting and fishing in forests may be for subsistence of Aboriginal peoples (Kant 2000). Needless to say, such a pattern simply opens the door to even more possible discontinuities.

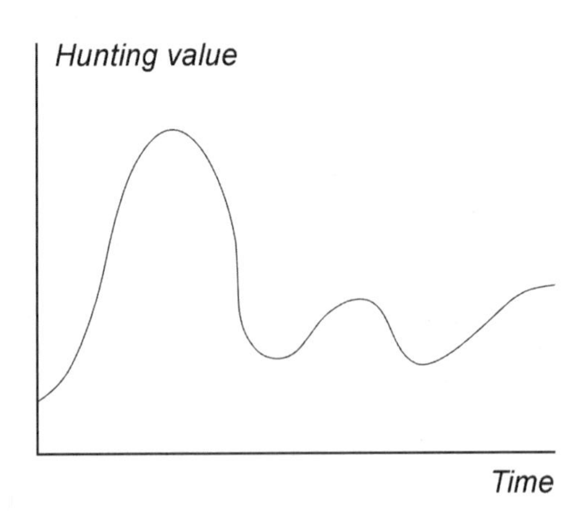

FIGURE 10.8 Virginia deciduous forest hunting amenity.

Although this chapter has no further detailed analysis of the various time patterns of potential nontimber amenities in forests, the list of such amenities that can (or should) be accounted for is quite long. They include a variety of biodiversity amenities (Perrings et al. 1995), carbon sequestration (Alig, Adams, and McCarl 1998), and reduced flooding and soil erosion (Plantinga and Wu 2003). By and large, most amenities tend to favor longer rotation periods for the majority of forests.

Probably the most dramatic discontinuities in forestry management involve stability-resilience trade-offs (Holling 1973). I noted earlier the case of pest management in regard to the spruce budworms, in which efforts to control the budworms in the late stages of their cycle by spraying generally only aggravates their eventual population explosion and the resulting collapse of the spruce forest, which leads to a pattern of hysteresis cycles (Holling 1965, 1986, 1988). Holling (1986) has documented several other cases where similar patterns arise.

Somewhat similar is the problem facing forestry managers regarding fire management. Longstanding policy in the U.S. national forests was to suppress almost all fires vigorously, even those naturally occurring due to lightning strikes. However, it is now understood that this policy achieved short-term stability at the risk of long-term lack of resilience because underbrush buildup can set up a forest for a truly devastating large-scale fire, as happened in Yellowstone National Park in 1988. Muradian (2001) argues that the relationship between fire frequency and vegetative density is one of multiple states, which supports the idea of the possibility of catastrophic dynamics, as we have seen.

In cases of fire, in addition to the vegetation issue, there are also implications for preservation of endangered species, with some doing best in the wake of fires at midrange successional stages, such as the Eastern Bristlebird in the United States (Pyke, Saillard, and Smith 1995). C. Clark and Mangel offer an analysis of this sort of case (2000, 176–181). Figure 10.9 depicts the time path of an endangered population after a fire. They argue that from the perspective of maximizing such an endangered species, controlled fires should be set when the population is the largest, which may entail relatively frequent fires. Of course, controlled fires can get out of control, which has also been a problem in recent years.

Finally, the issue of the size of harvest cuts is another matter of great controversy with respect to forests. In general, timber companies prefer to do large-size clear-cuts because they are less expensive in simple economic terms. However, aside from externalities such as flooding, soil erosion, and plain unsightliness, such clear-cuts have implications for endangered species management as well. Thus, it has been argued that nonlinear relations exist between largest patch size of habitat and the general degree of habitat destruction or fragmentation. The larger the harvest cut, the more likely the resulting habitat patch size may be smaller. More particularly, it is argued that with anything smaller than a certain habitat patch size, there is a collapse of the fragile species population (Bascompte and Solé 1996; Muradian 2001; for broader perspectives on species extinction dynamics, see C. Clark 1973, Solé and Manrubia 1996, and Newman and Palmer 2003). The implications of this potential for population collapse, in conjunction with the declining costs of harvesting larger-size cuts, are depicted in figure 10.10. The case for somewhat smaller cuts (but not too small) is clear.

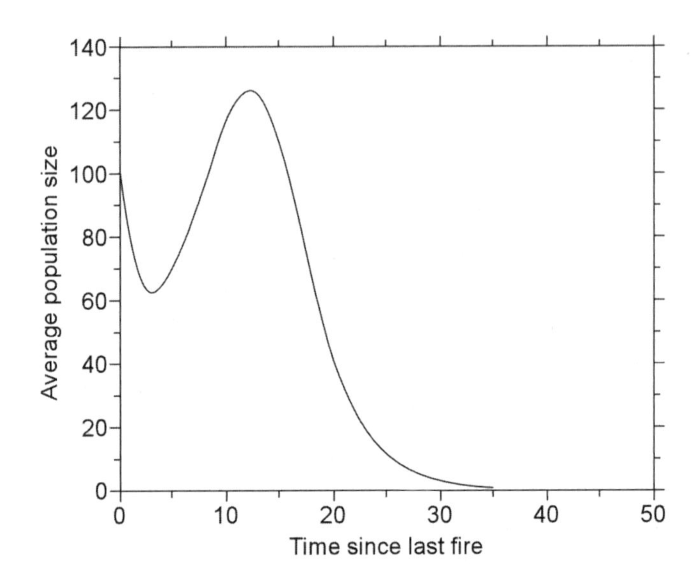

FIGURE 10.9 Average population path.

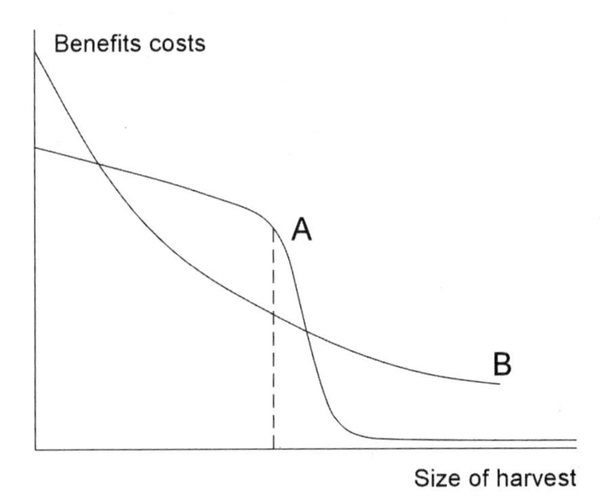

FIGURE 10.10 Harvest cut and habitat damage. The A curve represents the population able to be maintained as a function of the size of the harvest cut, and the B line represents the average costs of the cut.

HIERARCHY AND DISCONTINUITY

One of the deeper and more unresolved problems is how discontinuous dynamics operate within the context of ecologic-economic hierarchies. Considerable study of this problem has been done from the strictly ecological perspective (T. Allen and Starr 1982; O'Neill et al. 1986; Holling 1992), and Simon (1962) provided a more general foundation. A standard argument has been that higher levels constrain lower levels, but that lower levels do not constrain higher levels. However, this proposition has come into question in more recent years. Inspired by Diener and Poston's (1984) analysis of the "revolt of the slaved variables," Rosser and colleagues (1994) studied how events at lower levels might trigger changes at higher levels, especially discontinuous ones, a view also advocated by Gunderson and colleagues (2002). They also drew on Nicolis's (1986) theory of hierarchical entrainment to show how the "anagenetic moment" might arise, where a new level of hierarchy can emerge in a self-organizing manner out of lower levels.

Aoki (1996) has taken a related approach. He draws on Brock's (1993) mean field theory to study how externalities in a hierarchical system can bring about higher-level coherences and emergent structures. Fixed points in the coarse graining or aggregation of microunits are associated with sudden structural changes in the dynamical hierarchical system. In this model, sequences of phase transitions can arise as sequences of clusters of equilibria. A key to stability and a lack of such general hierarchical restructurings may be the existence of critical levels of a hierarchy whose stability ensures a broader order in the system, much like the role played by a keystone species in a more general ecosystem (Vandermeer and Maruca 1998).

Within combined ecologic-economic hierarchies, questions of management involve assignment of property rights and the careful assessment of appropriate policies related to the appropriate level. Rosser (1995) argues that property rights must be assigned to the relevant level of the combined hierarchy in order to facilitate appropriate policies. The importance of this appropriate assignment is seen by considering situations where it has not been done. Thus, Wilson and colleagues (1999) have shown how attempting to manage from too high a level of scale in fisheries can lead to overfishing of crucial local stocks as well as to their destruction, with broader implications at multiple levels for the ecosystem in question. This case reminds us again that social and economic organization by those most knowledgeable and in touch with a situation or system is often the best management path. The breakdown of traditional management of common property through interventions by people and institutions at higher levels—those not in touch with the system—has a long and tragic history. Likewise, attempting to manage a higher level from a lower level can be ultimately frustrating, as the ongoing failure to grapple seriously with global climatic change reminds us.

In ecologic-economic systems, there are many situations in which dynamic discontinuities appear. Many of these situations are of an undesired sort, especially when they involve the collapse or even the extinction of a species or the more general destruction of an ecosystem. Although I have noted a large variety of cases where such discontinuities occur, I have focused most discussion on the problems of fisheries and forests, both of them under broad and serious pressure from mismanagement by human beings. It is somewhat ironic that such discontinuities can arise when humans are trying to account for broader sets of concerns and phenomena than merely for simple economics, but certainly it is sometimes the emphasis on the latter that can be the source of problems. In any case, we must deal with the broader concerns. Hence, a more vigorous effort needs to be made to reform institutions and decision-making systems to ensure that awareness of critical thresholds of a damaging sort are properly understood and accounted for. There really is no excuse anymore for the heedless destruction of ecologic-economic systems.

11

THE ECOLOGICAL SIGNIFICANCE OF DISCONTINUITIES IN BODY-MASS DISTRIBUTIONS

Jennifer J. Skillen and Brian A. Maurer

THE BODY sizes of species within an assemblage are not evenly distributed along a continuum of possible body sizes (see references within Holling 1992; Havlicek and Carpenter 2001). Rather, these distributions typically contain significant discontinuities that indicate species of certain size ranges do not exist within the assemblage (Holling 1992). Discontinuous body-size distributions appear to be relatively common in natural assemblages, although the position and number of discontinuities vary among ecosystems and taxa (Schwinghamer 1981; Holling 1992; Allen, Forys, and Holling 1999; Havlicek and Carpenter 2001). This variation in the location and number of discontinuities implies that ecosystems vary in their ability to support organisms of particular sizes and may reflect the heterogeneity of environmental structure (Holling 1992).

The size of an organism determines the manner in which it interacts with its environment and therefore what constitutes available resources for it (Holling 1992; Milne et al. 1992). The combination of the environment's structure and a particular-size organism's ability to use and respond to that structure will influence population dynamics (Milne et al. 1992; Cuddington and Yodzis 2002). This combination ultimately determines which species can have self-sustaining populations in any given geographic region and will therefore contribute to the body-size frequency distribution of that assemblage. Holling (1992) proposed that a small number of biotic and abiotic processes, each of which operates at a particular spatial and temporal scale, govern the structure of terrestrial landscapes. These scale-specific processes create discontinuously structured landscapes. Organisms can interact with their environment only at a single scale at any given time, although some species

We thank Craig Allen for his assistance with the identification of discontinuities, the National Center for Ecological Synthesis and Analysis Working Group on Body Size in Ecology and Paleoecology for data, and Bert Skillen and Genny Nesslage for helpful comments on the manuscript.

may have the ability to shift their interactions between multiple scales (Holling et al. 1996; Allen and Saunders 2002). Using this approach, Holling (1992) suggested that organisms living in a discontinuously distributed landscape will exhibit attributes, such as range size or body size, that are also discontinuously distributed due to the hierarchical structure of resources in their environment.

In this chapter, we examine the relationship between the characteristics of landscapes across eastern North America and the avian species assemblages within them. We expect that a species assemblage within an ecosystem will have identifiable discontinuities in its body-size spectrum. Because discontinuities in the body-size spectrum represent body sizes for which suitable resources are not accessible in that landscape (Holling et al. 1996), species that are declining or are present in very low numbers in that landscape should be found close to the discontinuities in the size spectrum. Human impacts on landscapes result in changes in landscape structure and therefore in resources. We also expect that the pattern of discontinuities in the body-size spectrum will respond to the alterations in landscape structure caused by human activities. It has also been suggested that these changes will cause species turnover in an ecosystem, beginning with species occurring close to discontinuities in the size spectrum because these areas are "zones of crisis and opportunity" (Allen, Forys, and Holling 1999, 120). If this is true, we expect that in ecosystems impacted by human activities, not only will declining species occur close to discontinuities, but nonindigenous species will also be found there.

THE SPECIES ASSEMBLAGE

In North America, birds are widespread in every ecosystem, and their characteristics are well known. North American ecosystems (north of Mexico) have been represented by "physiographic regions" delineated for the North American Breeding Bird Survey (BBS) (Bystrak 1981). Physiographic regions are delineated based on physical characteristics and land use. Therefore, each region represents a relatively uniform biotic area, as required by Holling's (1992) hypothesis. Of the seventy-three physiographic regions described by the BBS, only ten are sampled by more than one hundred BBS routes. Of these ten regions, we chose six located almost entirely within the eastern United States with which to examine size structure (fig. 11.1). BBS routes consist of fifty sampling points located 0.8 km apart along secondary roads. Surveys are conducted once per year during June, and all birds seen or heard at each sampling point are recorded (Robbins, Bystrak, and Geissler 1986). For this study, a species list was compiled for each physiographic region based on all observations of terrestrial birds made on each route in that region over the course of the BBS. Because these regions are relatively small compared to the

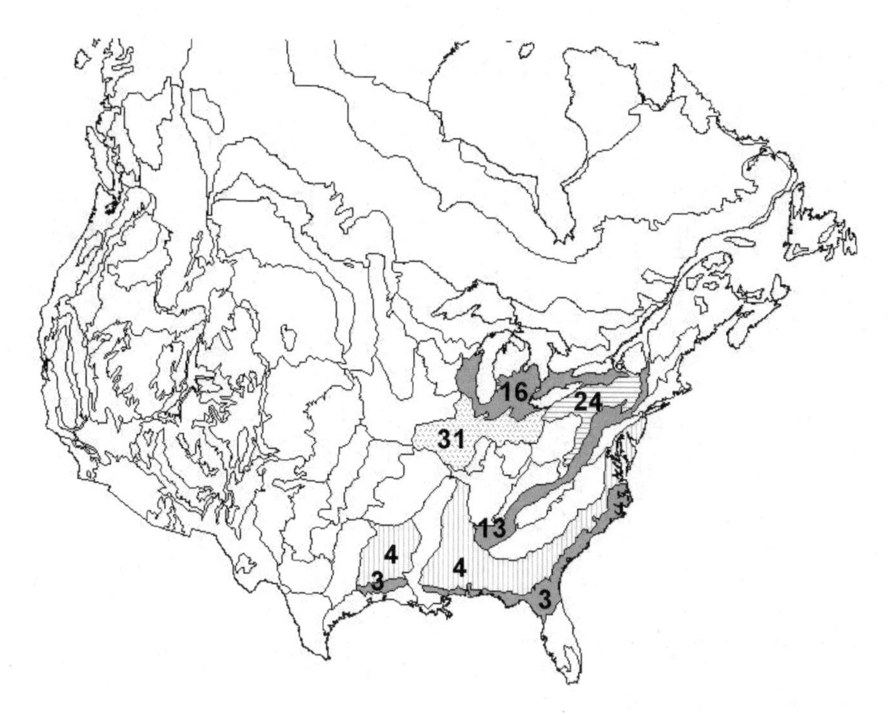

FIGURE 11.1 Physiographic regions delineated for the North American Breeding Bird Survey. 3—Coastal Flatwoods; 4—Upper Coastal Plain; 13—Ridge and Valley; 16—Great Lakes Plain; 24—Allegheny Plateau; 31—Till Plains (Bystrak 1981).

size of many bird species' geographic range and in many cases are adjacent to one another, there is overlap among the species assemblages of different regions.

The similarity among these species assemblages was assessed with a hierarchical agglomerative cluster analysis. The average linkage (Unweighted Pair Group Average, UMPGA) cluster analysis performed better than a Ward's cluster analysis (cophenetic correlations 0.87 and 0.79, respectively), though they differed only in the placement of region 31, the Till Plains. The dendrogram (fig. 11.2) shows that the species assemblages of the two coastal regions (3 and 4) are quite similar. The assemblages of the four noncoastal regions, especially the two mountainous regions (13 and 24), are also similar. Despite these similarities among regions, no two regions are inhabited by exactly the same complement of species.

Birds' body sizes can be measured with univariate and multivariate methods, but the best single descriptor is probably adult body mass (Rising and Somers 1989). Although using body mass as an estimator of overall size can be problematic because of seasonal and intraspecific variation (Gaston and Blackburn 2000), it also has distinct advantages. Body masses can be compared among

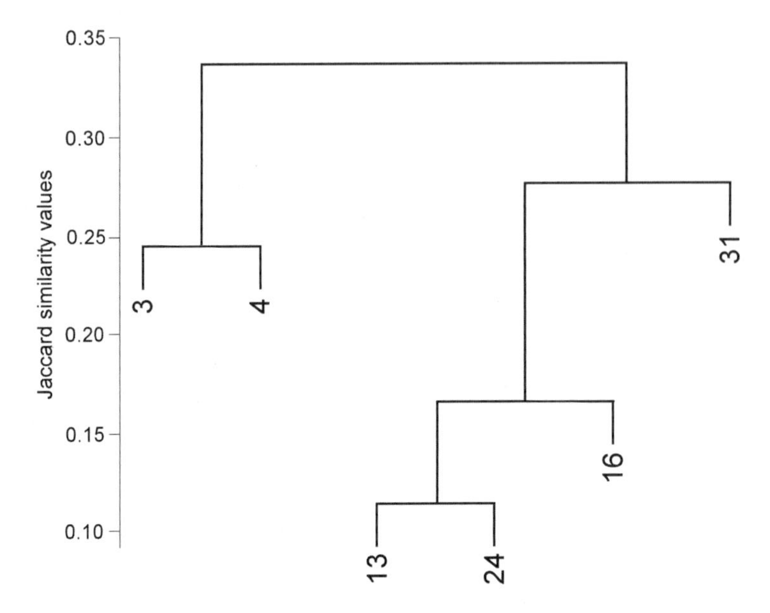

FIGURE 11.2 Average linkage cluster analysis for the species lists of the six physiographic regions based on Jaccard similarity values (cophenetic correlation = 0.87, agglomerative coefficient = 0.43). Regions are numbered as in figure 11.1.

taxa, unlike other size metrics such as wingspan or tarsus length. Because body mass can be used to estimate many other species characteristics, such as metabolic and reproductive rates (Peters 1983), it has been documented for many of the world's bird species. In our analyses, we use the body masses provided by Dunning (1993).

The six selected regions have experienced varying degrees of human modification and have varying native vegetation. Both the Allegheny Plateau (region 24) and the Ridge and Valley (13) ecosystems consist primarily of upland deciduous forest. The Coastal Flatwoods (3) is a varied ecosystem containing sand dunes, maritime forest, marsh, bottomland hardwoods, and pine. The slightly more interior Upper Coastal Plain region (4) is predominantly a mixture of upland pine forest and bottomland hardwoods. The Great Lakes Plain (16) is dominated by glacial topography and contains broadleaf forest, oak savanna, and prairie interspersed in the landscape. The Till Plains (31) are a mosaic of tallgrass prairie, savanna, and forest.

The species list for each region was assessed to determine which species were nonindigenous or declining (table 11.1). Nonindigenous species were defined as those introduced by humans from some other locality, such as the European Starling *(Sturnus vulgaris)*, or, in eastern North America, the House Finch *(Carpodacus mexicanus)*. We use the term *declining* to refer to those species that are

TABLE 11.1 Summary of Species Lists for the Six Physiographic Regions Used to Test for Discontinuity and Nonrandom Association of Nonindigenous and Declining Species in Terms of Body-Mass Pattern

REGION	NONINDIGENOUS SPECIES	DECLINING NATIVE SPECIES	TOTAL SPECIES
COASTAL FLATWOODS	6	18	130
UPPER COASTAL PLAIN	9	52	163
RIDGE AND VALLEY	6	64	158
GREAT LAKES PLAIN	6	43	158
ALLEGHENY PLATEAU	5	44	157
TILL PLAINS	7	18	129

at risk of extinction or local extirpation. In some cases, the abundance of a species may actually be increasing due to conservation efforts, though its numbers are still below historical levels. We identified declining species by consulting lists of protected species for each state that intersected a physiographic region. Species in the three highest-risk categories (state endangered, state threatened, and species of special concern) were considered to be declining. If a state's natural-resource agency did not maintain a list of state-protected species, then we obtained the information from the state's natural-heritage program. Because each physiographic region is intersected by multiple states, it was possible, though not typical, for a species to be defined as declining within the region even though it was considered to be at risk by only one state in the region. Every state has a different process for determining at-risk species; each of these processes is also influenced by local politics. As a result, our definition of a declining species may not accurately reflect the biological status of a species in all cases. However, the use of state lists is a practical method of describing species status. In addition, for each ecosystem the state lists used to generate the regional list of declining species were remarkably similar.

IDENTIFICATION AND MEASUREMENT OF DISCONTINUITIES

Discontinuities in the body-mass spectrum can be identified with several methods (Holling and Allen 2002). Each method has its strengths and weaknesses, so criticism can be leveled at any particular method (e.g., Manly 1996; Siemann and Brown 1999; see also chap. 9 in this volume). Confidence in the accuracy of identified discontinuities is increased if multiple methods produce the same result. Consequently, we used six methods to identify the discontinuities in the avian body-size spectrums. We used only the native species of an ecosystem in these analyses in order to determine where discontinuities exist in the recent historical fauna of a region (Allen, Forys, and Holling 1999).

HOLLING BODY-MASS DIFFERENCE INDEX

Holling (1992) described an index to measure how similar a species is to those species closest to it in size:

$$\text{Holling Body-Mass Difference Index (BMDI)} = (Mn+1-Mn-1)/(Mn)k,$$

where Mn is the body mass (in grams) of the nth species in order of increasing size and k is a constant that detrends the data. Holling found this constant to be taxon specific, equivalent to 1.3 for birds and 1.1 for mammals. A large value in the BMDI indicates that a species is quite different from those ranked closest to it in size. In order to determine how large an index value must be to represent a true discontinuity in the distribution, a criterion line is placed at the average index value plus one standard error (Holling 1992). Discontinuities in the distribution are identified by those index values that are greater than the criterion (fig. 11.3a).

SIEMANN AND BROWN INDEX

Siemann and Brown (1999) also developed a body-mass difference index:

$$\text{Siemann and Brown Index (SBI)} = \log_{10}(Mn+1/Mn)$$

They argue that this index is a more precise metric than the BMDI and that it requires no a posteriori defined exponent to correct the data.

ROBUST INDICES

Because of the concern that the beginning and ending values in each body-mass spectrum were producing artifactually high index values that would result in the identification of false discontinuities, we also used a robust version of the two indices. We visually identified exceptionally high index values at each end of the distribution, removed them from the calculation of the average index value, and then recalculated the criterion line for determining if a discontinuity was significant (fig. 11.3b).

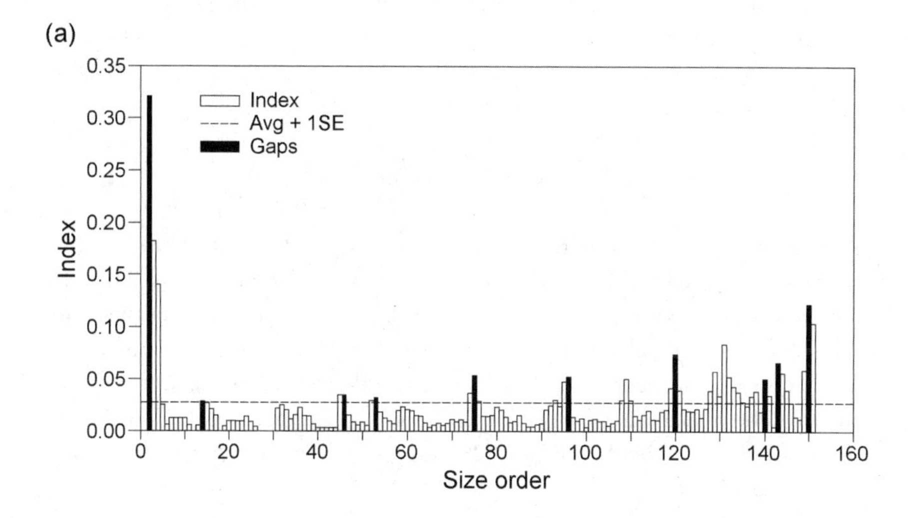

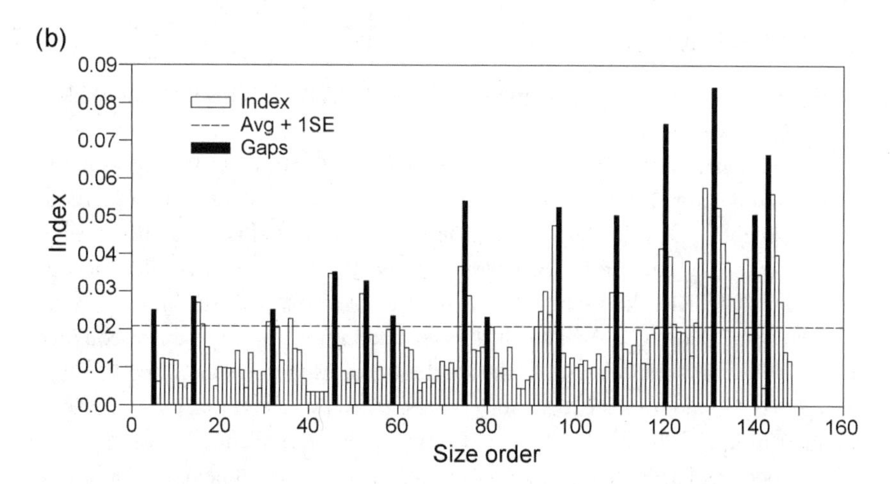

FIGURE 11.3 *(a)* Size-ordered Holling Body-Mass Difference Index (BMDI) values for the Ridge and Valley ecosystem. Criterion line has been placed at the average native species index value, plus 1 standard error (SE). *(b)* Robust size-ordered BMDI values for the Ridge and Valley ecosystem. The three smallest and three largest species have been removed and the criterion line readjusted.

GAP RARITY INDEX

The clump analysis Gap Rarity Index (GRI) (Restrepo, Renjifo, and Marples 1997) uses the GRI statistic, which is the probability that the observed discontinuities in the body-size spectrum occur by chance alone. The GRI statistic was calculated by comparing the actual data with a null distribution. The null distribution was generated by using log-transformed body masses to estimate a continuous unimodal kernel distribution (Silverman 1981, 1986). This null distribution was then sampled repeatedly, and an absolute discontinuity value was calculated for each species in simulation:

$$dn = \log_{10}(Mn + 1) - \log_{10}(Mn)$$

The distribution of the differences for the *n*th-largest species from the simulations was compared to the actual value obtained from the original data. The GRI statistic for each species in the actual assemblage is the proportion of the simulated discontinuity values that were smaller than the observed discontinuity value. The significance of each GRI value was then determined by testing the null hypothesis that the value was drawn from a continuous distribution with an alpha of less than 0.05. Significant GRI values represent discontinuities in the body-size distribution.

CLUSTER ANALYSIS

A hierarchical cluster analysis using PROC CLUSTER (SAS Institute 1999) identified aggregations in the body-mass spectrum. We used Ward's minimum variance clustering method (Ward 1963), based on a within-cluster least-squares criterion, because of its high performance compared to other clustering methods (Milligan 1981). Determining the appropriate number of clusters is difficult, but can be aided by looking for consensus among the three criteria calculated by PROC CLUSTER: the pseudo F-statistic, the pseudo t^2 statistic, and the cubic clustering criterion (CCC). The clusters formed in this way represent a sequence of clusters—from one containing the smallest species to one containing the largest species. Boundaries between clusters are formed by species that are next to one another in ranked body size. A discontinuity can then be defined as existing between species adjacent to one another in ranked body size but belonging to different clusters.

Discontinuities identified by the BMDI and the SBI, as well as their robust versions, were compared. Consensus discontinuities were defined as those identified by a majority (at least three) of these four methods (table 11.2) and were used in further analyses. For three ecosystems, these consensus discontinuities also were compared with the results from the GRI and the cluster analysis. In general, the six methods were quite consistent. Other studies (Holling et al. 1996; Holling and Allen 2002) have found that several additional methods identify the same discontinuities in a data set, although we did not use them here. These methods include the split moving-window index (Cornelius and Reynolds 1991), the Silverman Difference Index (Silverman 1986), a kernel estimator (Chambers et al. 1983; Silverman 1986), and classification and regression tree analyses (J. Clark and Pregibon 1992). The structure of body-mass distributions can be described in terms of the distance of each species in an assemblage to the nearest discontinuity (e.g., fig. 11.4). To determine these distances, species that defined the edges of a discontinuity were assigned a value of zero. Any species that occurred in a discontinuity was also given a value of zero. Values for all other species were calculated as the absolute value of the difference between the mass of that species and the mass of the closest discontinuity edge. Because the untransformed body-size distribution was positively skewed, the distribution of distances to nearest discontinuity

TABLE 11.2 Gaps Identified by Six Methods for the Coastal Flatwoods Data Set

BODY-SIZE RANGE (g)	HOLLING INDEX (BMDI)	ROBUST HOLLING INDEX (BMDI)	SIEMANN-BROWN INDEX	ROBUST SIEMANN-BROWN INDEX	GAP RARITY INDEX	CLUSTER ANALYSIS	CONSENSUS GAPS (g)
3–8	Y	N	Y	N	Y	Y	
8–8.5	N	N	N	Y	N	N	
8.8–9.4	Y	Y	N	Y	N	N	8.8–9.4
11–11.9	Y	Y	N	N	N	N	
13.2–14.1	N	Y	N	Y	Y	Y	
15–15.9	N	N	N	Y	N	N	
17–18	Y	Y	N	Y	N	N	17–18
21.6–23.6	Y	Y	Y	Y	Y	Y	21.6–23.6
38.7–41.7	Y	Y	N	Y	N*	N	38.7–41.7
55.3–61.5	Y	Y	Y	Y	Y	Y	55.3–61.5
86.8–102	Y	Y	Y	Y	Y	Y	86.8–102
245–300	Y	Y	Y	Y	Y	Y	245–300
632–1,028	Y	Y	Y	Y	Y	Y	632–1,028
2,172–4,130	Y	N	Y	N	N	N	

Note: Y = gap confirmed by mthod at this location.
Sources: Consensus gaps are based only on the Holling Body-Mass Difference Index (BMDI) and the Siemann and Brown Index and their robust versions.

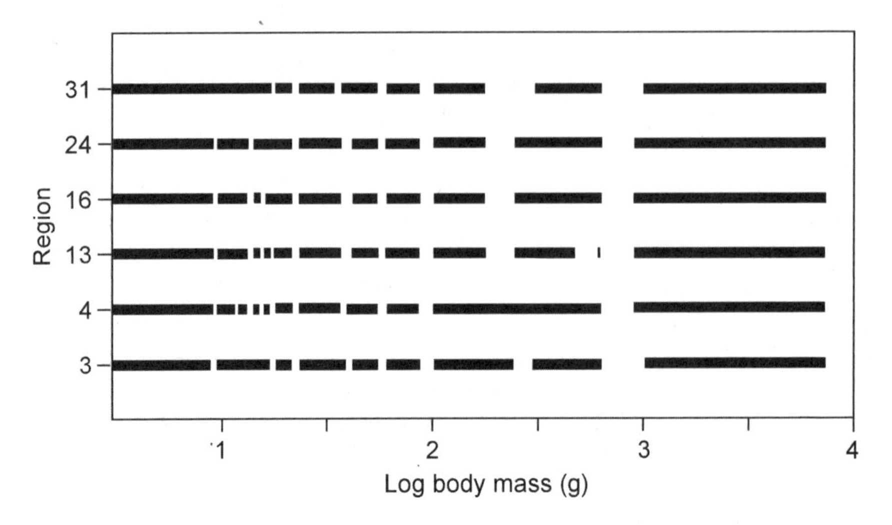

FIGURE 11.4 Location of discontinuities in the body-mass spectra for the six physiographic regions studied. Regions are numbered as in figure 11.1.

was also positively skewed. Before the distance to nearest discontinuity metric was log transformed, a small constant of one-sixth was added to each value to allow the use of zero distances (Tukey 1977).

THE PATTERNS

Our expectations for the patterns of the body-mass spectra were partially fulfilled. In each of the six physiographic regions, discontinuities in the body-mass spectra were identified with several methods, providing further support for Holling's (1992) hypothesis. Data from other taxa, locations, and scales have also been shown to support this hypothesis—birds and mammals of North American boreal forests and the short-grass prairies of Alberta (Holling 1992); birds, mammals, and herpetofauna of the Everglades; cave bats of Mexico; birds and mammals of Mediterranean-climate Australia (Allen, Forys, and Holling 1999; Allen and Saunders 2002); and marine sediment assemblages (Raffaelli et al. 2000; but see also Manly 1996, Siemann and Brown 1999, and Leaper et al. 2001).

DISCONTINUITY STRUCTURE

The overall pattern of discontinuities is shown in figure 11.4. The smallest species in each body-mass spectrum was the Ruby-throated Hummingbird

(Archilochus colubris) at the left of the graph, and the largest species was the Wild Turkey *(Melleagris gallopavo)*. There were consistently more discontinuities at smaller body sizes than at large body sizes, which reflects the larger number of small species in each assemblage. With a few exceptions, discontinuities were located at the same point in the body-mass spectrum, regardless of ecosystem. In the two regions with fewer discontinuities, the Coastal Flatwoods (3) and the Till Plains (31), lost discontinuities were those at the smaller body sizes. This observation illustrates two points. First, at a large spatial scale (represented by large body sizes), the discontinuity structure should be similar across ecosystems within a biome because the underlying large-scale geomorphological processes will also be similar. Discontinuities at smaller spatial scales would be expected to be similar across forested ecosystems within a biome because the small-scale structure of leaves and twigs is similar regardless of tree species. However, the small-scale structure would be expected to be different between forested and nonforested ecosystems. The Coastal Flatwoods and Till Plains regions may have fewer discontinuities at small body sizes because the forest component of these ecosystems differs from this component of the other four systems. In addition, a change in the discontinuity structure suggests that there has been a change in the complexity of the landscape (Holling et al. 1996). Because of differences in vegetation composition and the impact of human activities, the landscapes of the Coastal Flatwoods and Till Plains may have less complexity in their structure and therefore fewer discontinuities in their body-size spectra.

DECLINING SPECIES

For the six regions, there was a slight trend for the declining species to be closer to discontinuities than the other native species (an average of 52% of the declining species were closer to discontinuities than the median distance to the nearest discontinuity). The same pattern was also found for declining species in the Everglades (Allen, Forys, and Holling 1999). More than half (forty-two out of seventy-two) of the declining species found in multiple ecosystems had the same distance to the nearest discontinuity, regardless of ecosystem (fig. 11.5). The different distance measurements for thirty of the declining species reflect the slight changes in discontinuity structure from region to region (fig. 11.4).

NONINDIGENOUS SPECIES

Unlike the study by Allen, Forys, and Holling (1999), which also investigated the arrangement of invasive and declining species in the body-size spectrum, this study does not associate nonindigenous species with the discontinuities in the body-mass

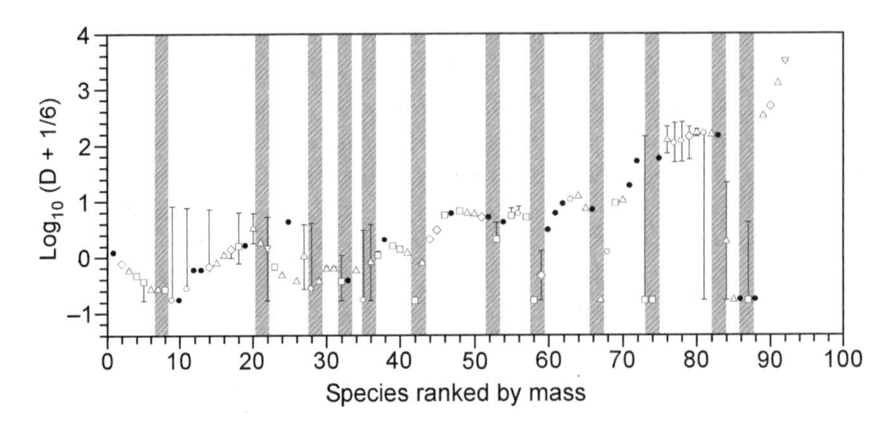

FIGURE 11.5 Median distance to nearest discontinuity for declining species across all six ecosystems studied. Error bars represent the range of values for those species whose distance to nearest gap varied across ecosystems. ● Species occurring in only one region ($n = 21$). △ Species occurring in two regions ($n = 30$). □ Species occurring in three regions ($n = 20$). ◊ Species occurring in four regions ($n = 9$). ○ Species occurring in five regions ($n = 10$). ▽ Species occurring in six regions ($n = 2$). Shaded vertical columns represent approximate locations of discontinuities in the body-size spectra.

distribution. Cumulative distribution functions of the distances to nearest discontinuities for each of the regions showed that an average of 72% (a range of 43% to 83%) of the nonindigenous species were farther away from discontinuities than the median distance (fig. 11.6). It is also apparent from these cumulative distribution plots that, as with the declining species, the occurrence of nonindigenous species in relation to the discontinuity structure was consistent from ecosystem to ecosystem. The House Finch *(Carpodacus mexicanus)* is widely distributed in all six ecosystems and was consistently located at the edge of a discontinuity in each distribution. The House Sparrow *(Passer domesticus)* and the European Starling *(Sturnus vulgaris)* are also widely distributed in all regions, but were always found near the median distance to the nearest discontinuity. The Eurasian Collared-Dove *(Streptopelia decaocta)*, Ringed Turtle-Dove *(S. risoria)*, Rock Dove *(Columba livia)*, Gray Partridge *(Perdix perdix)*, and Ring-necked Pheasant *(Phasianus colchicus)* had the largest distance to the nearest discontinuity in each region.

BODY SIZE

The positively skewed distribution of body sizes in each ecosystem means that there are far more small species than large species. Similarly, the pattern of discontinuity occurrence in the distributions (fig. 11.4) shows that most dis-

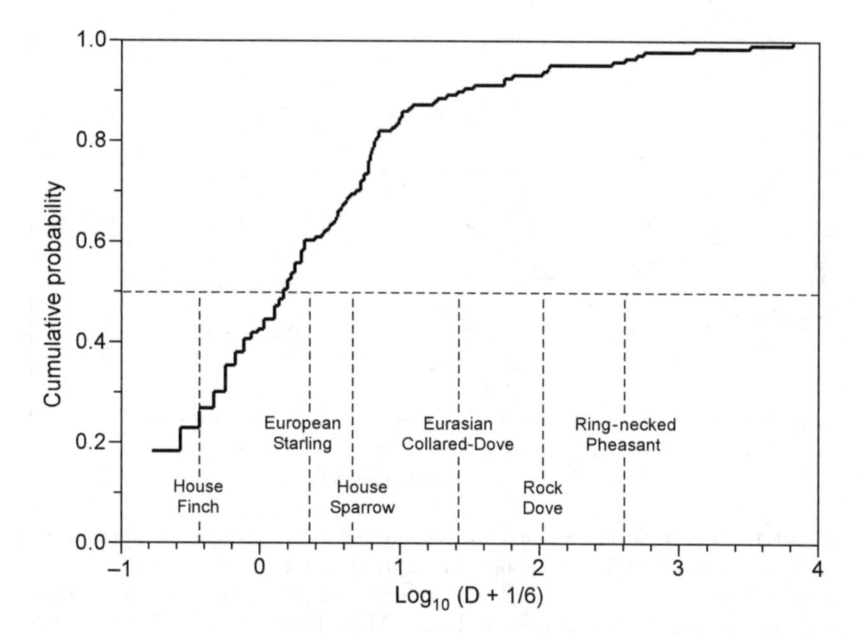

FIGURE 11.6 Empirical distribution function of distance to nearest discontinuity for all species in the Ridge and Valley ecosystem. The horizontal line denotes the median probability. Five of the six nonindigenous species have a distance to nearest discontinuity that is greater than the median distance for this region.

continuities occurred at small body sizes. As a result of these patterns, large-bodied species were usually farther from the nearest discontinuity than were small-bodied species. This difference is evident in a plot of the average distance to nearest discontinuity for nonindigenous species (fig. 11.7) and in a similar plot for declining species (fig. 11.8). However, it is worthwhile to examine those species that appear to be outliers and to consider what may make them unique.

The plot of average distance to nearest discontinuity for nonindigenous species (fig. 11.7) shows that based on its body size, the Monk Parakeet *(Myiopsitta monachus)* was located closer to a discontinuity than expected in the one region in which it occurs. In fact, this species was actually located within a discontinuity. Unlike the other nonindigenous species, which are ground gleaners (Ehrlich, Dobkin, and Wheye 1988), the Monk Parakeet can utilize a variety of foraging strategies (South and Pruett-Jones 2000). The House Sparrow *(Passer domesticus)* is widely distributed in all six ecosystems and was always found near the median distance to the nearest discontinuity. However, its congener of similar size, the Eurasian Tree Sparrow *(Passer montanus)*, was found within a

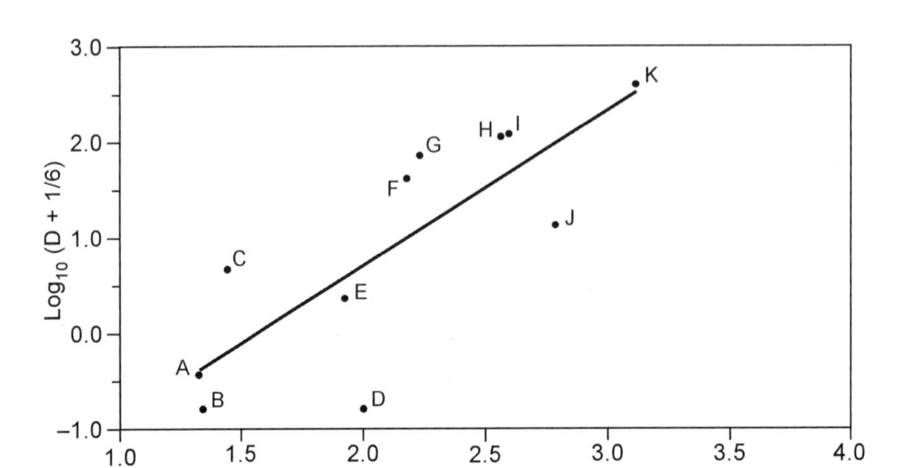

FIGURE 11.7 Average distance to nearest discontinuity for nonindigenous species ($n = 11$) across all six ecosystems with least-squares regression line ($r^2 = 0.65$). Bird names and (in parentheses) number of regions of occurrence: A—House Finch (6), B—Eurasian Tree Sparrow (1), C—House Sparrow (6), D—Monk Parakeet (1), E—European Starling (6), F—Eurasian Collared-Dove (3), G—Ringed Turtle-Dove (2), H—Rock Dove (6), I—Gray Partridge (2), J— Chukar (1), K—Ring-necked Pheasant (5).

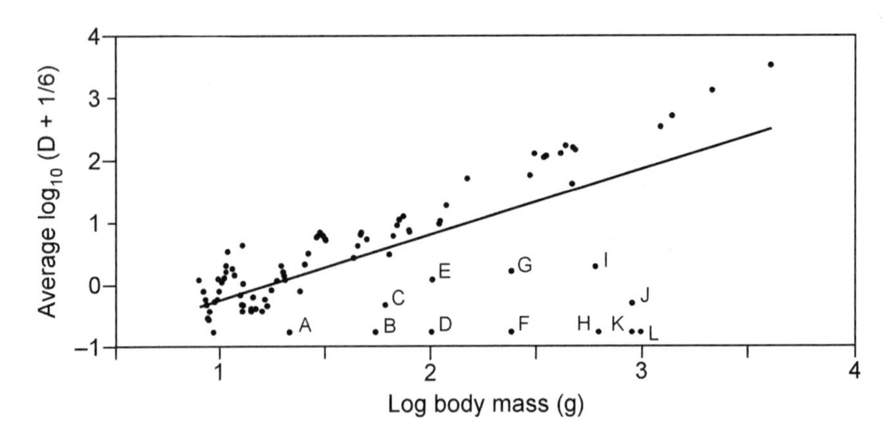

FIGURE 11.8 Average distance to nearest discontinuity for declining species ($n = 92$) across all six ecosystems with least-squares regression line ($r^2 = 0.55$). Species with unusually low distances to nearest discontinuity and declining are: A—Cliff Swallow (3), B—Whip-poor-will (3), C—Common Nighthawk (4), D—Eastern Meadowlark (2), E—Sharp-shinned Hawk (5), F—Long-eared Owl (3), G—Mississippi Kite (3), H—Barred Owl (2), I—Peregrine Falcon (2), J—Northern Goshawk (3), K—Swainson's Hawk (1), L—Greater Prairie-Chicken (1).

discontinuity. Although the Eurasian Tree Sparrow has been quite successful in its region of introduction, it has not expanded its range in North America since its arrival in 1870.

The trend of increasing average distance to nearest discontinuity with increasing body size is even clearer for the more numerous declining species (fig. 11.8). However, some species were much closer to discontinuities than expected for their body size. Interestingly, the opposite pattern does not appear. Many of these species defined the edges of a discontinuity in all regions in which they occur. These species include the Swainson's Hawk *(Buteo swainsoni)* and the Greater Prairie-Chicken *(Tympanuchus cupido)*, which occur in only one region, as well as the Cliff Swallow *(Petrochelidon pyrrhonota)*, Whip-poor-will *(Caprimulgus vociferous)*, Eastern Meadowlark *(Sturnella magna)*, Long-eared Owl *(Asio otus)*, and the Barred Owl *(Strix varia)*, which are declining in multiple regions. Other species, such as the Common Nighthawk *(Chordeiles minor)*, Red-shouldered Hawk *(Buteo lineatus)*, Peregrine Falcon *(Falco peregrinus)*, and Northern Goshawk *(Accipiter gentilis)* defined discontinuity edges in some ecosystems, but not in others.

REGION-SPECIES INTERACTIONS

A combined plot of the cumulative distribution functions of the distance to nearest discontinuity for all ecosystems (fig. 11.9) shows that most ecosystems have a very similar probability structure. However, the Till Plains were strikingly different. In this region, there was relatively little probability that a nonindigenous species would occur only a short distance from a discontinuity, but the probability increased rapidly at distances greater than the median distance to the nearest discontinuity. The Upper Coastal Plain appeared to be somewhat intermediate between the Till Plains and the other ecosystems. There was a significant interaction between region and species type (nonindigenous or declining; maximum likelihood contingency table analysis, $X^2 = 11.42$, $df = 5$, $P = 0.04$). In general, a small proportion of the nonindigenous species in each ecosystem was found at a distance less than the median distance to nearest discontinuity, and a relatively high proportion of declining species was found at a distance less than the median (fig. 11.10). However, this relationship was reversed in the Till Plains.

Our analysis of the body-mass distributions of terrestrial breeding birds across six ecosystems of eastern North America resulted in several clear patterns. Discontinuities in the body-mass spectra were clearly present in all six ecosystems. The distribution of these discontinuities in each body-mass spectrum was similar across ecosystems. Declining species were generally closer to discontinuities than expected, whereas nonindigenous species were generally farther from

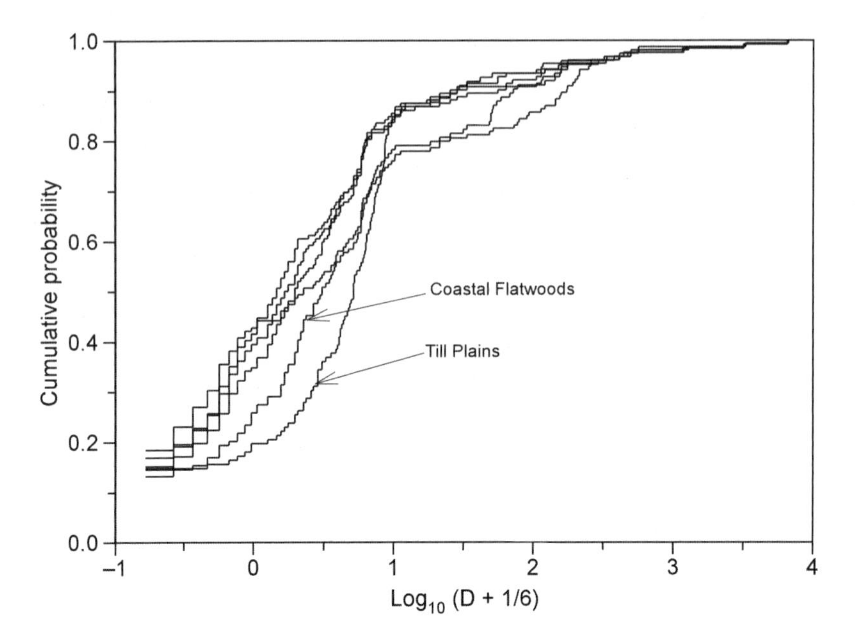

FIGURE 11.9 Empirical distribution functions of distance to nearest discontinuity for all six ecosystems. These distributions are significantly different (Kolmogorov-Smirnov statistic = 0.11, $n = 856$, $P < 0.005$).

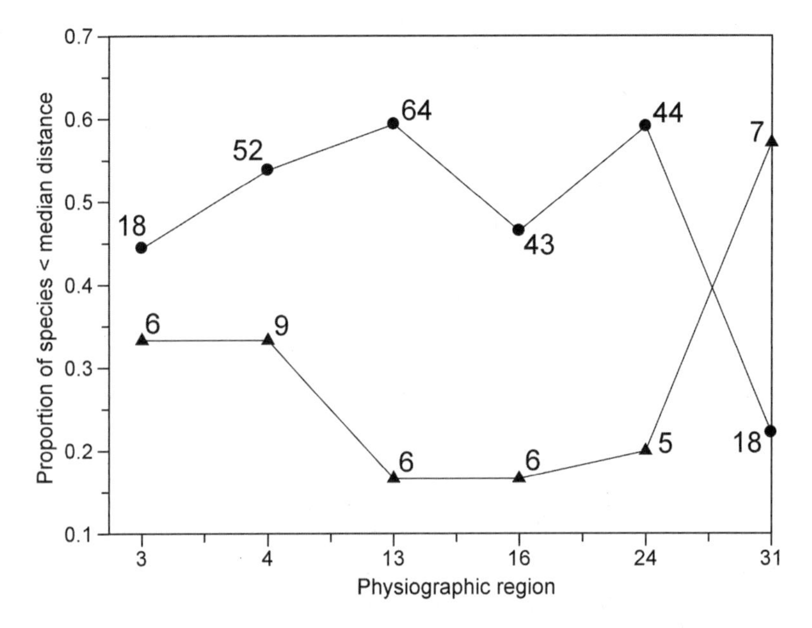

FIGURE 11.10 Proportion of nonindigenous (▲) and declining (●) species in each ecosystem that occur at less than the median distance to nearest discontinuity. Regions listed on the x axis are numbered as in figure 11.1.

discontinuities than expected. Body size generally increased with greater distance to nearest discontinuity. The structure of these discontinuities has an ecological significance, evidenced by the fact that there are consistent patterns across ecosystems and clear differences between successful invasive species and declining native species in an ecosystem.

MECHANISMS UNDERLYING DISCONTINUITIES

With so much environmental heterogeneity present among these six ecosystems, it is difficult to envision a simple mechanism underlying the discontinuities. Yet the discontinuities in the assemblages are nonrandomly located, and the species interact with these discontinuities in a nonrandom manner, which suggests that there must be some common underlying mechanism. We propose that this mechanism consists of the identifiable effects that the template of landscape structure has on population dynamics.

THE LANDSCAPE TEMPLATE

The composition of any landscape is a result of the past and present co-occurrence of biotic and abiotic elements in a certain spatial configuration. Climate, geology, topography, competition, predation, disturbance, and succession all interact to influence an ecosystem's components. A landscape's structure is described by the geometries of the biologic and geologic components (Turner, Gardner, and O'Neill 2001; Noon and Dale 2002). The presence and configuration of certain elements in a landscape will influence the characteristics of the ecosystem that exists in that landscape (Wiens 1995). The landscape's structure is scaled in a hierarchical manner by the temporal and spatial pattern of a few key processes (Holling 1992). For example, the dynamics of boreal forests and their insect defoliators are governed by four dominant cycles. Each cycle operates over a specific time period and spatial extent, such as the relatively fast, small-scale interaction between insects and needles, and the slower cycle of pest outbreaks that cause tree mortality over large areas (Holling 1992). This hierarchical structure is linked to the distribution of available resources. Any given resource exists at only one scale in the landscape. Landscapes that differ in structure will have differently scaled resources. This discontinuous structure forms a template upon which species' geographic ranges are overlaid. The template is both spatially and temporally dynamic because the types and abundances of resources change between locations and time periods.

The template of landscape structure is influenced by natural disturbances in the environment. Natural disturbances can be biotic (the grazing of large herbivores) or abiotic (fire, wind, drought). The spatial scale of disturbances varies over orders of magnitude; a raindrop may impact only one seedling (Begon,

Harper, and Townsend 1990), whereas a drought may affect a large geographic region. A small disturbance may create only an opening in the landscape (tree canopy gaps), but a large disturbance can reset a community's successional stage. Natural disturbances also vary temporally, with occurrences ranging from frequent (tidal cycles) to very rare (volcanic eruptions). Disturbances that happen often enough can invoke selection pressures on populations (Begon, Harper, and Townsend 1990). The impacts of disturbance can be assessed by measures of ecological resilience (Holling 1973).

Although landscapes have always been impacted by natural disturbances, anthropogenic disturbances are becoming increasingly common. Most changes in landscape pattern—including deforestation, conversion to agriculture, and urbanization—are currently a result of human activities (Turner, Gardner, and O'Neill 2001). Many of the anthropogenic changes to ecosystems are not unlike natural changes that have occurred over evolutionary time. Indeed, humans have influenced the structure of landscapes for millennia by changing the relative abundances of species (especially plants), altering the geographic ranges of species, aiding invasion by nonnative species, altering soil composition, and changing the distribution of cover types (Delcourt 1987). However, anthropogenic changes to ecosystems, including biological invasions and extinctions, are now occurring at a much faster pace than ever before (Vitousek et al. 1997).

Human impacts in North America, as in other continents, have been substantial. Forest clearing in the East began with the original Native American inhabitants and then increased dramatically after the arrival of European settlers. In the late 1800s, forest was being lost at a rate of 200,000 km^2 per year. With the abandonment of farms in the early 1900s and subsequent reforestation, rate of forest loss slowed to about 96,000 km^2 per year (Williams 1990). In the past three hundred years, North America has lost about 7% of its forests (Richards 1990), and almost every forest in the East has been cut at some point (Pimm and Askins 1995).

Human activities have affected in various ways the structures of the six ecosystems that we analyzed in this study (fig. 11.1). In many areas, not only has native vegetation been replaced by crops, but mixed crops have been replaced by single-crop monocultures, further reducing the landscape's structural heterogeneity (Warner 1994). The Till Plains ecosystem historically consisted of a mosaic of prairie, savanna, and forest, but almost 70% of this landscape has been converted to corn and soybean cropland (Fitzgerald, Herkert, and Brawn 2000). Approximately 15% of the region remains as some sort of pasture/grassland habitat, but the structure has been altered here, too. For example, the amount of land planted with a homogeneous monoculture of alfalfa has increased in recent years. Only a small amount of the original habitat is estimated to remain

(Fitzgerald, Herkert, and Brawn 2000). Land conversion for agriculture and urbanization, especially along the coastline, has led to the loss of a large fraction of native habitat in the Coastal Flatwoods region (Hunter, Peoples, and Collazo 2001). Most forested wetlands have been cut and fragmented, and intense development along the coast has led to the loss of maritime forest, dune, and marsh. The eastern portion of the Upper Coastal Plain has corn or soybeans across approximately one-third of its area. Most of the historically widespread southern pine habitat has been cut and converted to pine plantations or nonforest uses (Hunter, Peoples, and Collazo 2001). Almost the entire original savanna habitat in the Great Lakes Plain is gone, as is more than 80% of the forest. Croplands (corn, soybeans, hay, pasture, grains) are now the dominant land cover (Knutson et al. 2001).

Humans' influence on ecosystems has been dramatic and pervasive. Urbanization and conversion to agriculture homogenize both the structure of landscapes as well as the genetic, population, and species diversity of biota within them (Matson et al. 1997; Vitousek et al. 1997; August, Iverson, and Nugranad 2002). Agricultural practices have substantial effects on ecosystem processes, including the activities of soil biota, nitrogen and other nutrient cycling, and hydrological cycles (Matson et al. 1997). It is likely that as human activities continue to cause additions and deletions in regional species assemblages, ecosystem functioning will be changed. These changes may reduce communities' resistance to and resilience for future alterations to the landscape template (Chapin et al. 1997).

THE POPULATION RESPONSE

In a broad sense, the pattern of environmental structure across the landscape affects the response of species living in the landscape through variation in resources, which influences species' ability to survive and successfully maintain populations in a region. Species vary in their requirements and in their ability to utilize resources; each species has a unique niche (Hutchinson 1957). Therefore, each species will exhibit a unique response to the landscape. Any given landscape will be capable of supporting only a subset of species at any given time. In addition, species and even life stages within species differ in their capacity to resist and recover from disturbances. For example, species with high reproductive capabilities and broad environmental tolerances will often recover their former population levels relatively quickly after disturbances, whereas species with lower reproductive capabilities and narrower tolerances will be slower to recover. The species assemblage present in a region is the sum of the overlapping geographic ranges of species that occur in that region. A species' geographic range is located where the environment's conditions (biotic and abiotic) match that species' requirements, allowing it to maintain populations across the landscape. The geographic range is characterized by spatially autocorrelated gradients in population abundance and density that decrease from the center to the periphery of the range, though sometimes this gradient is cut off by a "hard" range boundary set by some physical barrier, competition, or predator-prey interactions (Brown 1984; Brown, Stevens, and Kaufman 1996; Curnutt, Pimm and Maurer 1996). Underlying these gradients is spatial autocorrelation in the environment's suitability for a species, which also decreases from the center to the periphery of the range (Brown 1984). The variability of population abundances is relatively low at the range's center and relatively high at its periphery (Pimm 1986; Curnutt, Pimm, and Maurer 1996). Combined with lower abundances and densities, this increased variability increases the likelihood of extinction for a population at the edge of its range. A range's boundary has been generally defined as the point at which populations cease to contribute to the next generation (Caughley et al. 1988); the boundary's exact position is determined by the interaction of population processes with the spatial and temporal variation in the environment (Brown, Stevens, and Kaufman 1996).

A species' geographic range commonly extends over more than one distinct ecosystem (99 out of 206 species were found in all six ecosystems in our study). A species' ability to maintain successful, stable populations within any given ecosystem is influenced by the relationship between the structure of the geographic range and the template of landscape structure. Each population within a species' geographic range has distinct demographic characteristics because each is influenced by a unique set of environmental factors. The dynamics of a local population can be described by the stochastic logistic population model:

$$\frac{1}{N} \cdot \frac{dN}{dt} = r - uN + \sigma z,$$

where N is population abundance, r is the maximum per capita rate of increase, u is the negative acceleration on population growth due to intraspecific competition, z is a white (Gaussian) noise random variable with mean zero and variance 1, and σ is a scale factor that represents the size of random fluctuations in the per capita rate of change (Dennis and Patil 1984; Dennis and Costantino 1988; Dennis 1989). Spatial variation in population dynamics can be modeled by assuming that the maximum per capita rate of increase (r) and the effects of intraspecific competition (u) are spatially explicit parameters. The influence of the environment causes these parameters to covary across the species' geographic range (Maurer and Taper 2002). At the center of the geographic range, where populations are typically abundant and stable, the maximum per capita rate of increase is high and decreases with distance from the range center. Conversely, the effect of intraspecific competition typically has the lowest effect on populations at the range center and increases with distance.

The biological complexity caused by the ecological individuality of species in a spatially and temporally varying environment can be described by combining the effects generated by hierarchical environmental structure and the effects on local population process generated by the position of populations within the species' geographic range. The relative unsuitability of resources near the periphery of a species' geographic range will influence the dynamics of populations occurring there; they will exhibit relatively low per capita rates of increase *(r)* and relatively large negative effects of intraspecific competition *(u)* (fig. 11.11). This scarcity of resources should be reflected in the structure of the body-size spectra for ecosystems near geographic range edges; a species in an ecosystem at the edge of its range will likely occur near a discontinuity in the body-size spectrum for that ecosystem.

Consider the hypothetical landscape depicted in figure 11.12b. The geographic range of species A is centered on Ecosystem 1. The environmental conditions for A are most suitable at the center of its range (Brown 1984), with appropriate resources to meet species requirements, and therefore populations of A in Ecosystem 1 will most likely be stable. This stability is depicted in figure 11.11, where populations of species A exhibit high maximum per capita rates of increase *(r)* and low effects of intraspecific competition *(u)* at the range center (Maurer and Taper 2002). As shown in figure 11.12, species A is relatively small and is located at the center of an aggregation in the body-size spectrum, far from discontinuities. As we move toward the boundary of the geographic range of species A, into Ecosystem 2, the environment's suitability and the availability of resources decrease. This decline is reflected in the position of species A in the body-mass spectrum; in Ecosystem 2, species A is located at the edge of a discontinuity. At the scale at which species A is able to interact with the landscape of Ecosystem 2, resources are scarce or inaccessible (Holling et al. 1996). Consequently, populations of species A in this region are likely to be relatively small and scattered, as compared with those in Ecosystem 1 (Brown 1984). They are also likely to be unstable and to exhibit low maximum per capita rates of increase, large effects of intraspecific competition, and relatively high variability (fig. 11.11) (Maurer and Taper 2002). Species B is larger than species A (note its relative position in the body-mass spectra) and interacts with the landscape differently. Because it has different requirements, its geographic range is centered on Ecosystem 2. Here, its populations are stable, with high per capita rates of increase and little negative effect from intraspecific competition (fig. 11.11). In Ecosystem 2, species B is located in the center of an aggregation in the

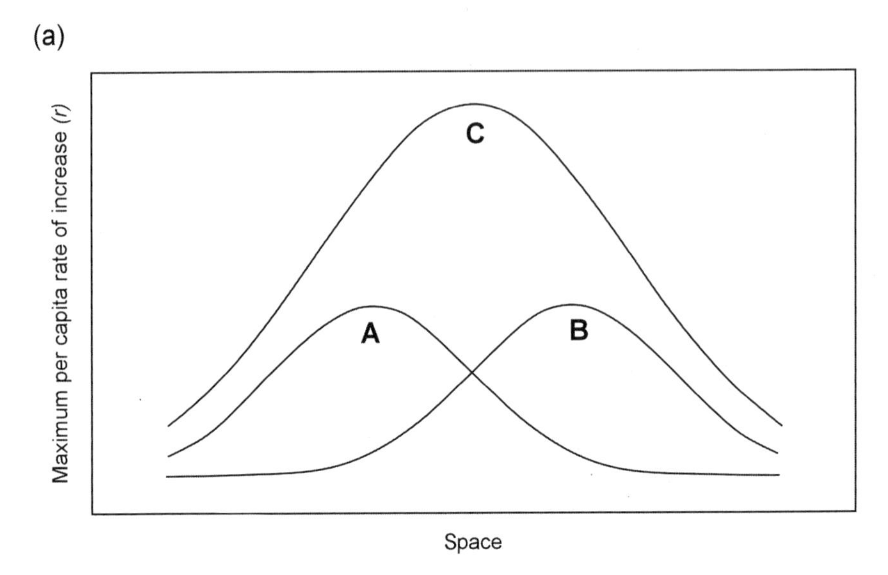

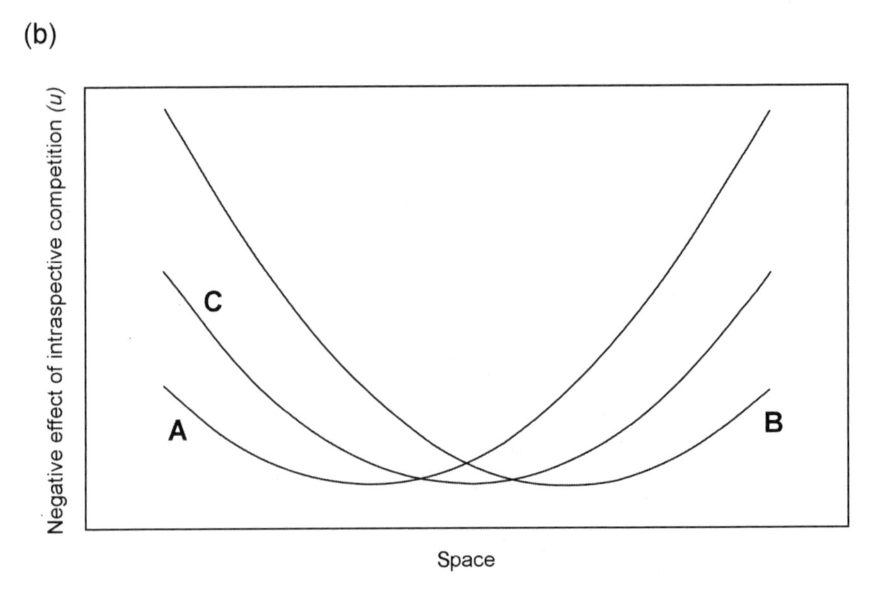

FIGURE 11.11 Schematic representation of *(a)* the variation of the maximum per capita rate of increase and *(b)* the negative effect of intraspecific competition across the geographic ranges of two narrowly distributed species (A and B) and one widely distributed species (C). The x axis corresponds to the horizontal transect in figure 11.12.

body-mass spectrum. Species B also occurs in Ecosystem 1, but its range boundary intersects this region. Consequently, its populations in this system are relatively unstable, with lower per capita rates of increase and greater negative effects of intraspecific competition (fig. 11.11). In Ecosystem 1, species B, at the edge of its geographic range (fig. 11.12a), is located at the edge of a discontinuity in the body-mass spectrum.

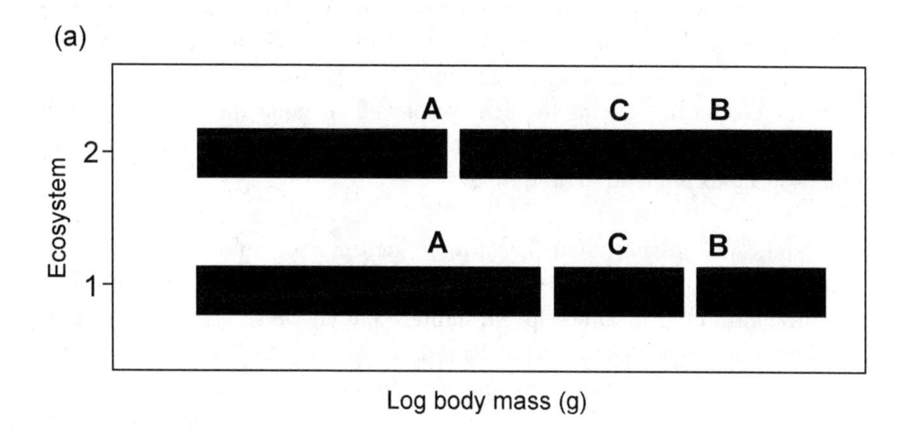

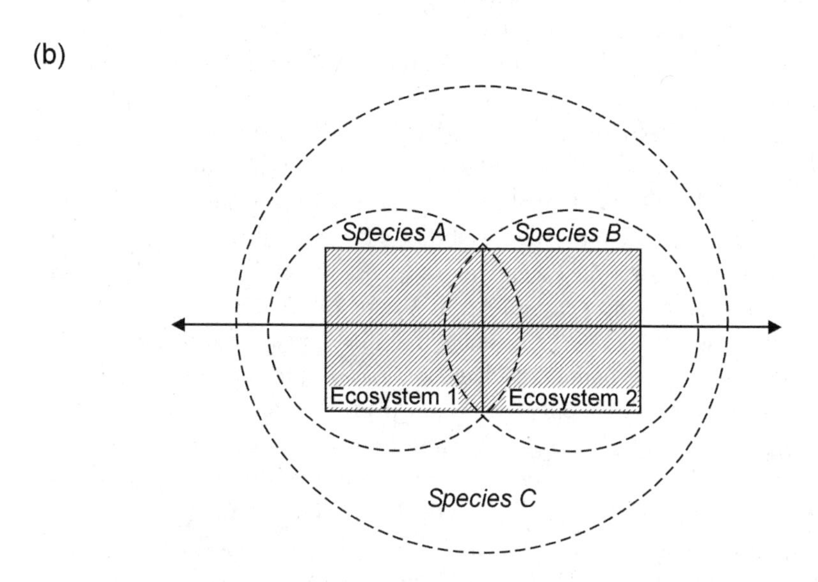

FIGURE 11.12 *(a)* A stylized representation of the discontinuous body-mass spectra for two ecosystems, with the position of species A, B, and C indicated. *(b)* Schematic representation of the geographic ranges of species A, B, and C in relation to two distinct ecosystems. The linear transect across the ecosystems corresponds to the *x* axis in figure 11.11.

Species C is intermediate in size to species A and B. However, it is a widely ranging species whose geographic range boundary intersects neither of the two ecosystems depicted here. Consequently, all populations of species C in these two ecosystems have sufficient resources available to them. Species C is not located near a discontinuity in either body-mass spectrum, but instead has stable populations throughout both regions, with high rates of increase and few negative effects from intraspecific competition. The actual population mechanism that determines the geographic range boundary for any species varies both spatially and temporally across populations, resulting in a varying pattern of extinctions and colonizations across the landscape (Maurer and Taper 2002). The overriding factor influencing the demography of any population at any point in time might be the maximum per capita rate of increase *(r)* or the negative effect of intraspecific competition *(u)* or both.

The landscape's discontinuous structure influences geographic variation in the abundances of species and therefore the composition of species assemblages. The discontinuities in landscape structure represent areas of either scarce or highly variable resources (Allen and Saunders 2002) where species can maintain only low-abundance populations and are typically at the edge of their geographic range. There is thus an element of risk for those species that occur at the edge of discontinuities in a body-mass spectrum. Even those that are not currently declining (as defined by state lists of protected species) or at the edge of their breeding range may be at risk in the future if there is sufficient change in the landscape.

However, the discontinuities in the body-size spectrum can also provide a zone of opportunity for introduced species (Allen, Forys, and Holling 1999; Allen 2006). Although we found that most nonindigenous species were farther from discontinuities than the median distance (fig. 11.6), the House Finch *(Carpodacus mexicanus)* was the exception in all six ecosystems. This bird is a recent and widespread invader of eastern North America (Elliott and Arbib 1953; Veit and Lewis 1996). It is important to note that it is a species whose range is still expanding. As such, its population dynamics may be interacting differently with the landscape than the population dynamics of a nonindigenous species that has stopped expanding its range (Maurer, Linder, and Gammon 2001). It is quite possible that in their native western landscapes, House Finches are not located at the edge of discontinuities in the body-mass spectra. Another logical, though currently untestable, conjecture is that perhaps when this species' geographic range stabilizes in the East, natural selection will move the species away from discontinuities by an increase or decrease in body size.

The alteration of a landscape through human activities will likely disrupt the spatial pattern of population processes that occur in that landscape. Disturbed landscapes commonly see an increase in the number of nonindigenous species (Vitousek et al. 1997; Western 2001). As with the House Finch across eastern North America, when these nonindigenous species are still in the invasive stage, they might be found close to discontinuities in the body-mass spectrum, where they can take advantage of opportunities present at that scale. Although declining species are typically located near discontinuities in the body-mass spectra, they are occasionally found far from discontinuities, as in the Till Plains (31) region (fig. 11.10). Two possible explanations can be offered for this anomalous result. First, the region may be adjusting to recent (and likely anthropogenic) environmental changes that have resulted in a scarcity of once common resources. The species assemblage is in transition, although the discontinuity structure of the region's body-mass spectrum has not yet changed. As the altered landscape causes additions and deletions to the species pool, the structure of the body-mass spectrum will change, and declining species will then display the typical pattern of placement close to the discontinuities. Second, those declining species that are located far from discontinuities may simply be exhibiting a lagged response to some previous change in the environment. However, this second scenario seems unlikely in light of the results from the six ecosystems analyzed to date (see figs. 11.8 and 11.10).

The scarcity of suitable resources for species at the edge of their geographic range affects their population dynamics (low maximum per capita rates of increase, significant effects from intraspecific competition) and their position in the body-mass spectrum for the ecosystem. In the body-mass spectrum, those species that are at the edge of their geographic range in that ecosystem will be located close to discontinuities. This interaction between local population processes, the discontinuous structure of the environment, and the structure of the geographic range means that for any ecosystem, the relative values of population demographic parameters can be predicted from descriptions of the ecosystem's body-mass spectrum and the geographic ranges of the species in the assemblage.

Holling (1992) outlined a research agenda for the analysis of cross-scale dynamics in ecosystems (see also Holling and Allen 2002) that should ultimately move us toward an understanding of how a specific hierarchical landscape structure leads to a specific discontinuous pattern of body sizes (Holling et al. 1996). The results from our examination of six ecosystems suggest that a landscape's structural complexity may be the key to understanding the discontinuity pattern of its species assemblage. The connection between the landscape's architecture and the species residing in that landscape has important implications for ecosystem management and conservation. Holling (1992; also Holling et al. 1996) suggested that this relationship might be used to form an "eco-assay." That is, the knowledge that a landscape is changing at a particular scale, due to either intentional or unintentional activities, can be used to predict which species in an assemblage are likely to be affected. The opposite is also true: the knowledge that species of a particular body size are being affected (e.g., population declines) can be used to predict the scale at which a landscape is changing (Lambert and Holling 1998).

12

CROSS-SCALE STRUCTURE AND THE GENERATION OF INNOVATION AND NOVELTY IN DISCONTINUOUS COMPLEX SYSTEMS

Craig R. Allen and Crawford S. Holling

ECOSYSTEMS ARE characterized and structured by interactions among biotic and abiotic processes operating at discrete scales. Within scales, these interactions reinforce one another to create persistent structures and patterns. Across scales, different patterns and processes dominate and are only loosely coupled with processes at higher or lower scales. Abiotic processes interacting with biotic elements produce loosely structured hierarchical systems with emergent qualities such as resilience. Reinforcement and inhibition among interacting processes drive this organization. The partitioning of process, structure, and function within and across scales provides resilience to complex systems and opportunities for the systems' elements—be they species or firms or cities.

The change in structure and pattern with scale in complex ecological systems provides different templates at different scales with which animals may interact. Within a range of scale, species strongly interact, and the result is a diversity of species' lifestyles—and thus functions—and both actual and potential competition among species is reduced (Peterson, Allen, and Holling 1998). Redundant function is present among species within a system, but coexistence of "redundant" species is facilitated when those "redundant" species live at different ecological scales, producing reinforcement of functions across scales. Changes in domains of scale are characterized by distinct scale breaks reflecting abrupt changes in pattern and structure. Heightened variability at the species, population, and community level has been observed at those scale breaks (Allen, Forys, and Holling 1999; Allen and Saunders 2002, 2006; chap. 11 in this volume; C. Allen, unpublished material), as indicated by phenomena such as invasion, extinction, nomadism, and migration—reflecting, we believe, heightened variability and increased opportunities for innovation.

Here we offer a conceptual framework that explores some of the forces creating innovation and novelty in complex systems. Our model is broadly applicable to ecological theory, including community ecology, resilience, restoration, and policy. Understanding the sources of variability and novelty may help us better understand complex systems. Characterizing the link between landscape change and the composition of species communities may help policymakers in their decision-making processes. Understanding complex phenomena such as invasions, migration, and nomadism may enable us to grasp better the structure of ecosystems and other complex systems and aid our attempts to cope with and mitigate these phenomena (in the case of invasions) and to better understand or predict them or both. Knowledge of how variability is related to system structure and how that relationship generates novelty may help us understand how resilience is generated. Finally, we hope that our ideas provide insight into the evolution of complex behaviors and entities. We begin with a brief description of panarchical organization in complex systems, describe a typology of novelty, and end with a discussion of how panarchical, discontinuous structure may generate novelty and innovation.

PANARCHIES IN COMPLEX SYSTEMS

An ecosystem is the product of nonlinear combinations of its component parts. As a complex adaptive system, it possesses emergent properties, such as resilience and a discontinuous structure that varies across scales. This structure has been recently described as a *panarchy* (Holling and Gunderson 2002), a nested set of adaptive cycles. An *adaptive cycle* (Holling and Gunderson 2002) describes the process of development and decay in a system (fig. 12.1). The initial stage of development is a rapid exploitation and garnering of resources by system components, of short duration, and has been termed the *r* stage or function. This stage is followed by a *k* stage or function, a stage of longer duration characterized by the accumulation of biomass or other system elements or energies and by increasing connectivity and rigidity. The latter leads to decreased resilience and eventual collapse. This stage of collapse, the omega (W), is rapid and unleashes the energy accumulated and stored during the *k* phase. Collapse during the omega phase is followed by reorganization during the alpha *(a)* phase, a relatively rapid period of assembly of components, analogous to the pioneer stage in ecosystems.

It is important to note that adaptive cycles do not exist in isolation. They operate over limited ranges of scale, whereas complex systems are characterized by multiscalar organization and dynamics. A panarchy is a nested set of adaptive cycles (fig. 12.2). Each cycle operates over a discrete range of scale in both time and space and is connected to adjacent levels (adaptive cycles). Unlike the top-down control envisioned in traditional hierarchies, connectivity between adaptive cycles in a panarchy can be from levels above or below.

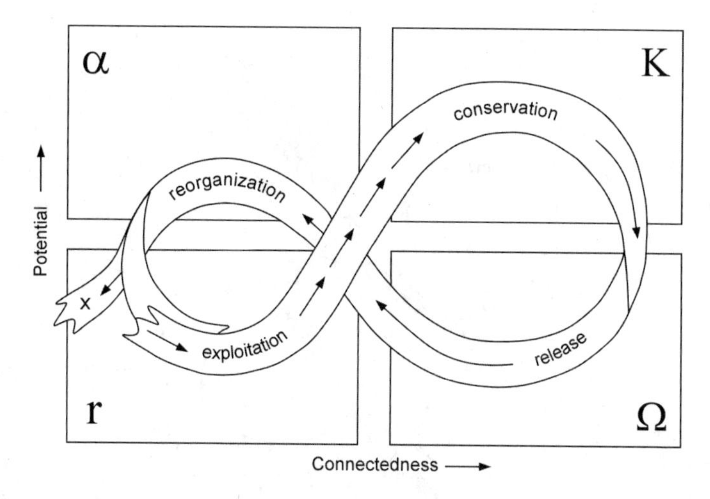

FIGURE 12.1 The adaptive cycle. A representation of the four ecosystem functions (*r*, *K*, *Ω*, *α*) and the flow of events among them. The arrows show the speed of the cycle, where short, closely spaced arrows indicate a slowly changing state and long arrows indicate a rapidly changing state. The cycle reflects changes in two properties: (1) *y* axis, the potential that is inherent in the accumulated resources of biomass and nutrients; (2) *x* axis, the degree of connectedness among controlling variables. Low connectedness is associated with diffuse elements loosely connected to each other whose behavior is dominated by outward relations and affected by outside variability. High connectedness is associated with aggregated elements whose behavior is dominated by inward relations among elements of the aggregates—relations that control or mediate the influence of external variability. The exit from the cycle indicated at the left of the figure suggests where the potential can leak away or where a change of state into a less-productive and less-organized system is likely. (Reprinted from Gunderson and Holling 2002 with permission of Island Press.)

ECOLOGICAL DYNAMICS, PANARCHY, AND DISCONTINUITIES

The structure of complex systems is strongly self-organizing and may be very conservative. This structure provides a system's core "memory." The components of complex systems such as ecosystems interact to create conservative structures in time and space. This conservativeness is important for humans because complex systems such as ecosystems often remain apparently stable, so we can count on reasonably predictable dynamics and the relatively constant provision of the ecological goods and services on which we depend. It is due in part to the interaction of biotic and abiotic elements. Animals interact with the ecological structures that provide a distribution of necessary resources such as food and shelter that they can successfully exploit in space and time. And in

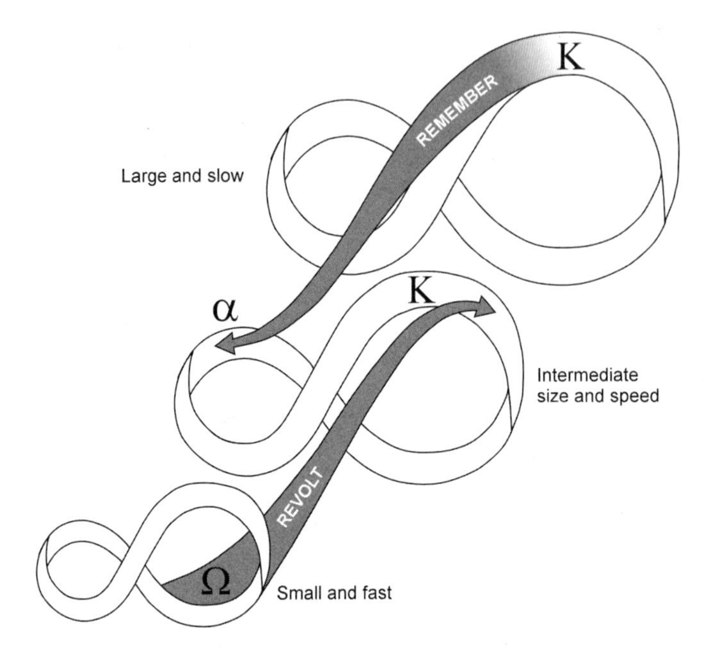

FIGURE 12.2 A panarchy. Three selected levels of a panarchy are illustrated to emphasize the two connections that are critical in creating and sustaining adaptive capability. One is the "revolt" connection, which can cause a critical change in one cycle to cascade up to a vulnerable stage in a larger and slower cycle. The other is the "remember" connection, which facilitates renewal by drawing on the potential that has been accumulated and stored in a larger, slower cycle. The number of levels in a panarchy varies, is usually rather small, and corresponds to levels of scale present in a system. (Excerpted from Gunderson and Holling 2002 with permission of Island Press.)

exploiting their environments, animals often change ecological structures in ways that are favorable for themselves. For example, large herbivores such as elephants alter the dynamics of succession (and competition among grasses, bushes, and trees) such that the habitat is, in some sense of the word, optimal for them. Self-organization involves other biotic system elements as well; for example, many grasses worldwide are pyrophyllic and are themselves highly flammable (Brooks et al. 2004). In the absence of fire, succession will often eliminate these grasslands from the ecosystems they occupy, but their presence encourages fire, which in turn favors their spread and excludes competitors. Strongly interacting species have been termed *keystone* (Mills, Soulé, and Doak 1993) or *driver* species. However, we note the absence of scale in such definitions. Because any keystone species (or process) affects a system over only a limited range of scales, it is a keystone only at that scale (unless it has truly cross-scale impacts, which is unlikely given current definitions of keystone species). Ecosystems are characterized by a rich array of scales, so there may be a rich array of keystone species present. Moreover, a keystone or driver species often has but a transient role; as system or community dynamics change, so does the identity of key species. Self-organization is important not only for the provision of goods and services to humans, but also because it means that understanding ecosystems and other complex systems is at least somewhat tractable. These systems' structures and functions, though dynamic through time, are largely constant and educible at human timescales. The exception, of course, is when the resilience of systems is exceeded and surprising transformations occur.

Because adaptive cycles and self-organization occur at discrete scales within a system, ecosystems and other systems are characterized by discontinuity. Such discontinuities have their roots in the separation between levels of a panarchy. Different adaptive cycles and different structuring processes are separated from one another by large gaps, often an order of magnitude or greater (Holling and Gunderson 2002). This separation has several important effects. First, as stated earlier, it means that variables within systems are distributed discontinuously. Second, it means that self-organizing interactions and processes, such as community-level interactions for animals, are compartmentalized by scale. Compartmentalization of systems along an axis of scale provides rich opportunities for experimentation within levels—which leads to the development of high levels of diversity within systems. This compartmentalization results in a distributional pattern of function where diversity is high within scales and repeated at different scales, a pattern that adds to the resilience of ecosystems (Peterson, Allen, and Holling 1998) and other complex systems (Garmestani et al. 2006). Finally, the presence of discrete scales of pattern and process in complex systems creates discontinuities, or scale breaks, between ranges of scales. Theory and recent empirical analysis suggests that these scale breaks generate and foster novelty and innovation as a result of the variable dynamics at these transitions.

VARIATION AT SCALE BREAKS

Discontinuities, or gaps, in variables in complex systems have been described as *scale breaks*, where highly variable and unpredictable behavior is expected (Allen, Forys, and Holling 1999). Examples of high variability at scale breaks in ecosystems include the success of invasive nonindigenous species, the failure or decline of native species, increased variability in a species' abundance in both space and time, and the presence of both migratory and nomadic species. The association of invasive species with scale breaks has been documented in a variety of ecosystems and for both birds and mammals. It has been most thoroughly investigated for the Everglades of Florida, where invasive species occur at the edge of body-mass aggregations (Allen, Forys, and Holling 1999). The association is significant because it links an independent biological attribute—invasiveness—with a particular location on a body-mass axis, at the edge of discontinuities. This association is strengthened by the finding that declining species, too, are associated with scale breaks and further by the demonstration, through analysis of successfully versus unsuccessfully introduced avian species in the Everglades, that introduction success is best predicted by distance from the edge of a discontinuity (Allen 2006). That is, an introduced species whose body mass places it close to the edge of a body-mass aggregation is more likely to become established than an introduced species whose body mass places it in the center of an aggregation. Other potential predictors of introduction success that are based on intrinsic or community-level hypotheses of introduction success do not work for that continental data set. It should be noted that despite a flurry of focus on invasive species over the past fifteen years, invasion biology is still a discipline in its infancy, largely because it lacks much ability to experiment and because inference is based mostly on positive cases (successful introductions or invasions); we hardly ever know what species were unsuccessful in becoming established, although there are exceptions. One of those exceptions is the south Florida avifauna, where we know with some surety the pool of unsuccessfully introduced species. The Everglades provides an additional clue to the link between ecological structure and novelty. Declining species compose about 25% of the fauna in three vertebrate taxa (Forys and Allen 1999), and nonindigenous species compose an additional 25%. Forys and Allen (2002) examined historic (no nonindigenous species), current (nonindigenous and declining species included), and hypothetical future (nonindigenous species included, declining species eliminated) body mass and functional patterns for mammals, birds, and herpetofauna of the Everglades. They found that neither the distribution of function nor overall body-mass pattern was substantially changed by the large species turnover in the time series studied. Body-mass pattern was conserved, and the large species turnover was limited mostly to areas of discontinuity.

Species invasions and extinctions represent turnover in animal community composition, but we also have documented other variability at scale breaks in animal communities. Migration and nomadism are annual or unpredictable periodic turnover; they are also novel and poorly understood behaviors. Both allow the exploitation of resources at a level that could not otherwise be achieved, at least not without some other novel innovative approach to secure resources that vary so strongly in both time and space. In Mediterranean-climate Australia near the city of Adelaide, the climate is highly variable, and the avifauna has among the highest—perhaps the highest—incidence of nomadism in the world. Among a suite of potential predictor variables in that region, nomadism is best modeled as a combination of body mass, nectivory, and proximity to discontinuity. Birds that are bigger and feed on nectar and that have body masses that place them closest to discontinuities are most likely to be nomadic (Allen and Saunders 2002, 2006; but see Woinarski 2006). Similarly, migrant birds have body masses that tend to place them at or close to discontinuities in body-mass distributions. We investigated the relationship between migrant species and body-mass distributions for the migrants that breed in the state of South Carolina in the United States, but that annually depart for the winter. We did the same for species that migrate from several places to Costa Rica (La Selva) during the winter, but that breed elsewhere. We found that in both South Carolina and Costa Rica, migrants tend to have body masses proximate to discontinuities. Interestingly, the South Carolina migrants that winter in Costa Rica, which have body masses close to scale breaks in the body-mass distribution of South Carolina resident birds, are not close to scale breaks in the body-mass distribution of Costa Rica birds (South Carolina migrants are a small subset of the migrant species present in Costa Rica). This finding suggests that existing near discontinuities and their variable resources represents an unexploited opportunity, but a risky one as well. It may be too risky to exploit "at the edge" on both the wintering and the summering grounds.

NOVELTY AND INNOVATION

Novelty and innovation are required for systems to remain dynamic and functioning. Without innovation and novelty, systems become overconnected and dynamically locked, so the capital therein is unavailable. Novelty and innovation are required to keep existing complex systems resilient and to create new structures and dynamics following system crashes. This is true in all complex systems, and the importance of novelty is recognized as much in the management and business world as it is in the scientific one—more so, perhaps.

Decades ago, writing of management hierarchies, Pierce and Delbecq (1977) described the organizational elements required for innovation. One of those elements is differentiation, which is necessary for the initiation of innovation. Differentiation is present in ecological systems in many forms—genotypically, phenotypically, and functionally—as well as in social and social-ecological systems. Decentralization is another important element described by Pierce and Delbecq. It is a key component of complex systems and related to Pierce and Delbecq's concept of stratification, which we interpret as levels within a hierarchy. Pierce and Delbecq also discuss contextual attributes of innovation, specifically environmental uncertainty, larger size, and age. Larger hierarchies have more opportunities for innovation, and older hierarchies are less open to innovation. Novelty is clearly important for the maintenance and health of a wide variety of systems—ecological, social, cultural, and combinations thereof. Novelty and innovation are also needed to sustain systems' adaptive capacity, above and beyond their maintenance functions, and to allow complex systems the latitude to "explore" alternative structures and dynamics—that is, to evolve. In the next section, we describe a typology of novelty as related to the structure of discontinuous, panarchically organized, complex adaptive systems.

A TYPOLOGY OF NOVELTY

The generation of novelty and innovation is a characteristic of dynamic systems at all levels. For example, in biological systems, novelty and innovation are generated at the genetic level through random processes of mutation, at the species level through the selective processes of evolution, at the community level as a result of assortment and changes in the species pool, and at the ecosystem level as a result of changes in key driving processes and self-organizing interactions.

One can organize the concept of novelty around fundamental similarities. We organize it into three kinds: background, incremental, and punctuated, each of which we discuss separately. All three kinds of novelty can be generated either locally or globally. Locally novel additions are new to that particular system. For example, the addition of an invasive species to an ecosystem adds novelty to that system, but the invasive species was in existence prior to its invasion of a new ecosystem. In contrast, globally novel additions did not exist previously and are new not only to the particular system within which they are generated or added, but also to the globe in general. Speciation within a system represents the addition of global novelty.

BACKGROUND

Novelty is generated as a result of complex systems' normal dynamics. In terms of panarchical, discontinuous structure in complex systems, we find novelty generated at scale breaks (between levels) as a result of the highly variable distribution and occurrence of resources in space and time, which in turn are reflected in high variability in the system's biotic components. This generation of novelty creates options for a system, maintains a system's adaptive capacity, and serves as a reservoir of potential functions that may be required following transformations or as normal system dynamics evolve. Such novelty is at the heart of resilience.

INCREMENTAL

Processes of self-organization among biotic and abiotic elements of complex systems such as ecosystems add complexity over time. Some of that complexity is added during the *r* and *k* stages of an adaptive cycle in the form of new connections, new functions, and new arrangements of elements. But during *r* stages, new levels (new adaptive cycles) can also be added. New layers to a hierarchy add new scales of opportunity for elements of complex systems such as species or firms. The addition of new levels of adaptive cycles may make a panarchical complex system more resilient and less prone to cross-scale collapse, (i.e., the collapse of all levels of a panarchy at once), or may make them increasingly vulnerable.

PUNCTUATED

When a complex system's resilience is exceeded, the system may collapse—all levels of a panarchy may experience the omega phase simultaneously. Novelty may be added to or introduced to a system during reorganization. Here, the novelty added may be local or global. Although this type of transformation and novelty generation is not likely to spawn globally novel elements, it does provide opportunities for globally novel arrangements of elements.

NOVELTY AND RESILIENCE

Both the addition of levels (or any other incremental novelty) to a panarchy and the generation of novelty at scale breaks (background) build resilience in systems. The novelty generated at scale breaks and, just as important, the potential for the generation of novelty at scale breaks build adaptive capacity in complex systems. Jain and Krishna (2002) document how dormant innovations in complex graph networks can take over system dynamics at times when other dominant elements become weak (i.e., when the system's resilience is diminished). Having a constant source of innovation and novelty is clearly important for systems, both following transformations and during their normal dynamics, if a system is not to become overconnected and overcapitalized. Without a continual source of novelty, complex systems such as ecosystems cannot be adaptive and dynamic. Scale breaks are a key source for such innovation. It should be noted, however, that innovation and novelty may be a destructive force as well. An invasive species, for example, can alter basic process and structure in an ecosystem and be a source for collapse and transformation. Thus, innovation and novelty may be a double-edged sword in some circumstances. In ecosystems, for example, "innovation" is both the prime source for recovery as well as a cause of major extinctions (Jain and Krishna 2002).

THE GENERATION OF NOVELTY AT SCALE BREAKS

The novelty generated at scale breaks, or those regions between adaptive cycles in a panarchy, is critical for complex systems, and these regions represent "novelty pumps," or regions of actual and potential production of innovation and novelty. As discussed earlier, highly variable phenomena are associated with discontinuities in animal body-mass distributions. These distributions reflect the scales of available structure in an ecosystem, which in turn reflect the system's panarchical structure (fig. 12.3). Thus, discontinuities in body-mass patterns reflect the location of scale breaks in ecological structure. A similar structure is also present in other complex systems, such as regional urban systems (Garmestani, Allen, and Bessey 2005; chap. 8 in this volume) and regional economic systems (Garmestani et al. 2006).

Phenomena such as invasion, extinction, nomadism, migration, and variability in abundance in animal communities reflect high variability, but they also represent the creation or insertion of novelty. Compared to native species, invasive species have subtly or grossly different ways of interacting with their environments,

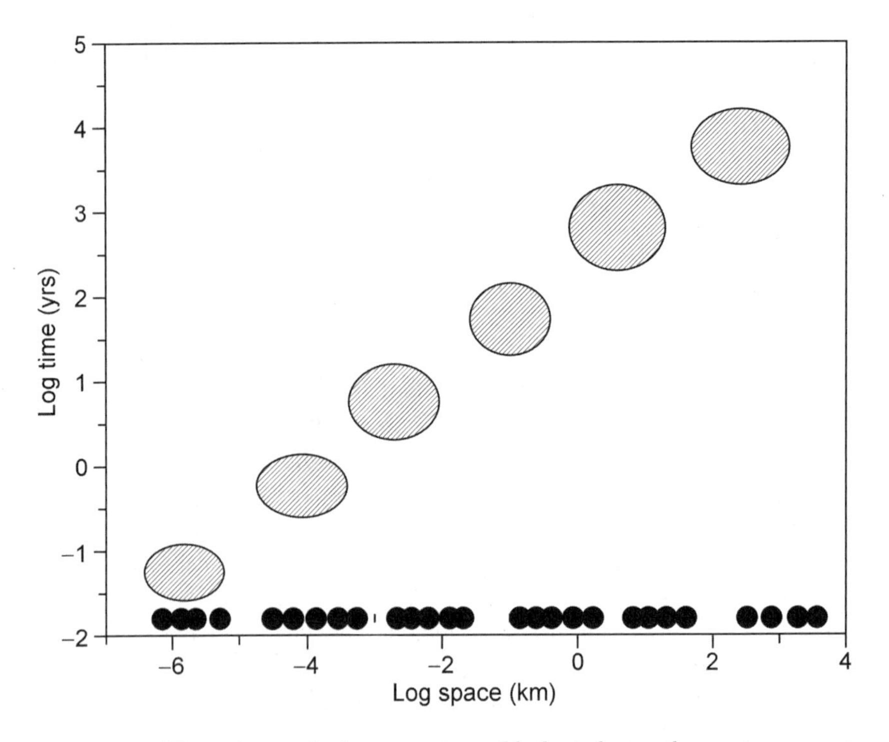

FIGURE 12.3 Discontinuous body-mass pattern (black circles on the *x* axis represent animal body masses along a body-mass axis) reflecting levels of scale (the hatched circular regions represent space-time domains of structures), which in turn reflect the underlying panarchy.

and their addition may reflect a system in transition (Allen, Forys, and Holling 1999). They may reinforce existing ecological organization (Forys and Allen 2002) and thus build resilience, or they may transform that organization.

The generation of novelty at scale breaks results, we believe, from high variability in resources (fig. 12.4). Although high variation in resource abundance and in location in space and time is a hardship for some species, witness declining species' propensity to have body masses proximate to discontinuities (Allen, Forys, and Holling 1999; chap. 11 in this volume); it is an opportunity for other species, which successfully invade these locations or which develop novel and innovative behaviors (fig. 12.5). However, a strategy that focuses on resources that are highly predictable, especially when the structure of a system is dynamic and where the location of scale breaks may shift over time, is probably not a good long-term strategy. However, it is the best strategy when the stable resources far from a discontinuity are effectively sequestered by others and niches are saturated.

Mutations and other novelties that occur as background have little chance of success in the center of an aggregation, far from discontinuities and scale breaks. At the center, resources are stable and thoroughly exploited. However, at discontinuities, the high fluctuation in resources is more likely to lead to success of random mutations. Species or strategies that are successful at the edge may over time migrate into the center of an aggregation, where resources are more stable and secure (fig. 12.6). This process is analogous to the Taxon Cycle in biogeography (Wilson 1961). Many of the major innovations in the history of life—such as the spread of novel metabolic activities in the first billion years of Earth's history, the spread of photosynthesis, the development of multicellular organisms,

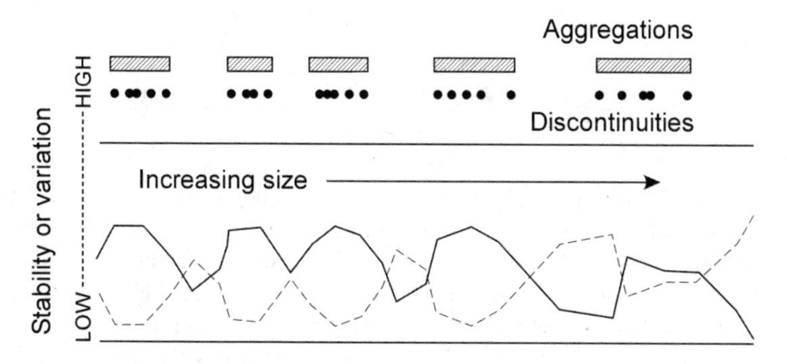

FIGURE 12.4 The relationship between discontinuities, variability, and stability. Body-size distributions of animals are characterized by aggregations and discontinuities (upper-graph circles represent species). Aggregations reflect domains of scale and discontinuities reflect scale breaks. Within domains of scale, resource variation is low (lower-graph dotted line). At scale breaks, resource variation is high. Stability (lower-graph solid line) is inversely related to variability and is high within scales and low at scale breaks.

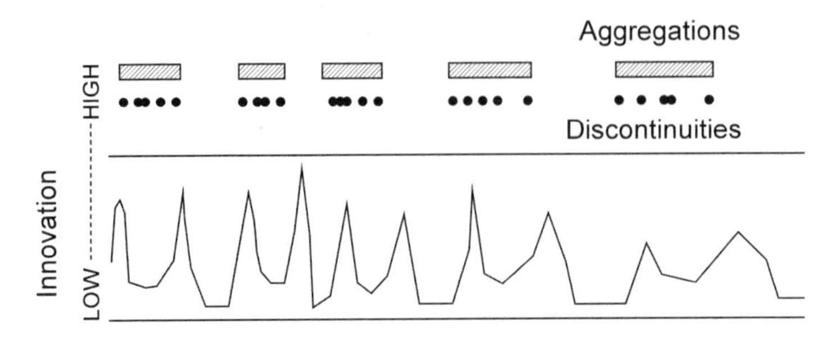

FIGURE 12.5 Novelty and innovation mirror discontinuities and resource variability in complex systems and are high at the edge of discontinuities.

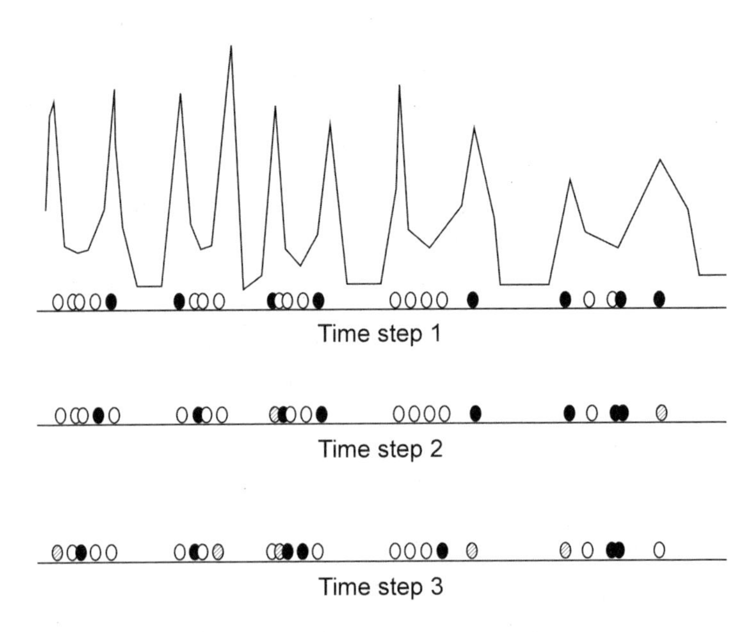

FIGURE 12.6 The novelty cycle in discontinuous panarchical systems is analogous to the taxon cycle described in Wilson 1961. Variability in resources (the line above the *x* axis that represents the discontinuous body-mass distribution of species) is highest at scale breaks, as in figure 12.5. In time step 1, novel behaviors or species (black) are added at scale breaks. As time progresses, time step 2, species' novel behaviors have "migrated" toward the center of body-mass aggregations, where resources are less variable. By step 3, this migration has been completed. In time steps 2 and 3, new novelty is being introduced at the edges (hatched). This diagram does not show the increased extinction rates also expected to be associated with scale breaks.

and the spread of life to terrestrial ecosystems—all share basic similarities. In each case, innovation was constructed through an evolutionary triad of challenge, potential, and opportunity. Each of these conditions is met within a panarchy and specifically at the discontinuities that define regions between levels—in other words, at scale breaks.

SPECIATION

Evolutionary innovation is the acquisition of novel morphologies and behaviors that open new niches, providing new ways to exploit the environment successfully. In ecosystems, an extreme case of the generation of novelty is represented by speciation. Scale breaks may offer opportunity for parapatric or sympatric speciation. However, the isolation necessary for speciation is not linear distance per se, as in parapatric speciation (Knapp and Mallet 2003), any more than it consists of geographic barriers. Rather, it is driven by the interaction with the environment at different ranges of scale. This interaction at different ranges of scale may strike some as unlikely. However, consider the extreme mammal example: least shrews and moose in a boreal ecosystem. These animals are allopatric, yet they interact little, if at all. The life span of a least shrew is less than two years, that of a moose more than twenty. The resources and structure with which a shrew interacts during the process of its life are, we believe, experienced simply as noise by the moose. Similarly, the structures and processes with which a moose interacts are so slow relative to a shrew's life span and step size that they are simply background to the shrew's tapestry. Life at discontinuities means constant adaptation to fluctuating resources. The high variability inherent at scale breaks and the more stable opportunities for resource acquisition at higher or lower scales provide ample opportunity for differentiation of lineages within a species—by habitat, by scale, by resource, or by a combination thereof. Because species at discontinuities are at the edge of an available range of scale, it is perhaps possible for these species to shift between scales (Allen and Saunders 2002). This possibility provides an opportunity for sympatric speciation based on scale. Moreover, the biological variability witnessed at scale breaks at the population and community level hints of potential individual phenotypic and genotypic variability as well.

Despite an enormous volume of papers on speciation from field, laboratory, and theoretical perspectives, the process leading to the creation of new species remains poorly understood (Sepkoski 1998). Although sympatric speciation is (may be) controversial for some researchers (Via 2001), habitat choice has been shown to produce assortive mating and sympatric speciation under disruptive selection (Rice 1984; Kondrashov and Mina 1986; Rice and Salt 1990), including selection of traits associated with competition or predation (Schluter 1996;

Via 2001). Sympatric speciation can arise from reproductive isolation associated with adaptation to alternative resources or habitats (Turelli, Barton, and Coyne 2001). Scale segregation may produce ecological speciation by isolating two populations based on their scale of habitat use; this speciation may be sympatric or parapatric. The rebound of species diversity (i.e., increased speciation rates) following mass extinctions (Sepkoski 1998) suggests that incipient species are always potential, but their "success" is greater following such events. Thus, a novelty pump, as it were, is extremely important in providing a source of new species following such events. Niche construction (lineages modifying their environments to construct their own ecological role and consequently to construct niches for other species) seems to be especially important following mass extinctions. Why? Panarchy theory provides a possible explanation. We cannot think of species as actors on an ecological or landscape stage, but rather should think of them as partners with their environment, traveling on a shared trajectory, where a change by the biotic elements or a change by the abiotic elements is equally important. Thus, mass extinctions either cause or reflect systems that have been transformed. The large loss of species means a disruption in self-organizing processes that are responsible for structure in complex panarchies. The loss of elements responsible for self-organizing processes with abiotic elements leads to the generation of new interrelationships, new self-organizing processes, and thus keystone, or niche-building, species.

APPLICATIONS

Novelty is in some sense just noise. At the molecular and other levels, novelty is being constantly created—and extinguished. Scale breaks provide a robust pump for the generation of novelty and thus are a key source of adaptive potential following transformation and adaptation of system elements such as species and of the systems themselves, regardless of transformative events.

Global climate change will bring about rapid transformations in the organization of the complex systems that we inhabit, rely upon, and create. The novelty pump inherent in the structure of complex panarchies will be necessary for adaptation to changing environments (for example, species that are already undergoing long-range migration or nomadic movements may be better able to cope with global climate change).

Species in the center of aggregations have "built" niches that are stable. Given the conservative nature of the structure of body-mass distributions (Havelcik and Carpenter 2001; Forys and Allen 2002) and the high degree of resilience in many systems, this strategy is wise for most circumstances. It is analogous to the *k* phase of the adaptive cycle. However, this strategy is apt to be "brittle" and vulnerable to failure when a system's resilience is exceeded and the system transforms or when other transformative events occur at one or more scales. The range of strategies with high variability found at scale breaks are likely to become dominant during such systemic crises.

Continuously pumping out novel solutions to current or potential challenges ensures a maximization of energy and resource use within a system without reorganizing the system, and it allows the system to be dynamic both in its internal structure and connectivity as well as in its relationship with other systems. Most of the novel solutions created by mutation and other sources do not have a challenge to respond to and thus are conceived of as failures. However, when challenges arise, it is critical to have a solution in waiting. Thus, novelty generation at scale breaks helps ensure that the evolutionary potential of both species and systems is maintained.

SYNTHESIS

Donald Ludwig

SOME OF the greatest scientific advances have originated with recognition that seemingly minor fluctuations and variations can reveal a pattern. The supreme example is Darwin's recognition that individual differences between members of the same species are not meaningless variations from an ideal type, but the basis for natural selection. This process was the foundation for Darwin's theory of evolution. A second example is Karl Jansky's recognition that some of the static that interfered with trans-Atlantic radio transmission originated from the center of the Milky Way galaxy. Jansky's insight led to the field of radio astronomy (see http://www.nrao.edu/whatisra/).

The present volume is a response to C. S. Holling's recognition in 1992 that there might be a pattern in the distribution of body sizes within taxonomic groups in a given ecosystem: so-called gaps and clumps. At first, there was little agreement on whether such patterns existed at all, on the criteria to detect such patterns, and on the significance to be attached to such patterns if detected. Allen, Forys, and Holling (1999) made a decisive advance when they showed that invading species and species in danger of extinction had a tendency to appear near the boundaries that separate clumps of body sizes within a taxon. Their main results are shown in figure 1.3 in this volume. The Santa Fe Institute sponsored a meeting in May 2001 to consider the controversy over the pattern of body sizes, and I interviewed some of the participants. Jim Brown felt that results of his study with Evan Siemann (Siemann and Brown 1999) had discredited Holling's idea. Nevertheless, after hearing Craig Allen's presentation of his work with Elizabeth Forys and Crawford Holling (Allen, Forys, and Holling 1999), Brown thought it would be worthwhile to test those ideas on data from the Breeding Bird Survey. Brian Maurer shared Brown's reservations, but nevertheless he actually carried out tests using those data. The results appear in Skillen and Maurer's contribution to this volume (chapter 11), which I discuss later. Shortly after the meeting in Santa Fe, Maurer wrote to me:

> The lack of mechanistic explanations and models for the distribution of gaps and lumps in ecosystems I think limits the usefulness of the lump/gap concept to a heuristic tool that can be used to illustrate why ecosystems are likely to be hierarchical. As yet, I have not seen many, if any, general, synthetic theories that can be used to predict the existence of hierarchical structure in ecosystems. I believe that part of the problem is that the tools that we have at our disposal to develop theories (e.g., logic, mathematics) are severely limited in their ability to describe extremely complicated systems of any kind. This often is brought home to me when I hear explanations couched in relatively simple mathematical constructions such as "equilibria" or "optima." Such ideas are attempts to describe in simplistic mathematical language the fact that many attributes of ecosystems seem to be fairly persistent in space and/or time. There was much of this kind of language being used at the meeting, and at least once I objected to use of it on the grounds that it was inadequate to describe the complexity that seems to exist in most ecological systems.

Maurer raised fundamental issues: Is it enough to recognize patterns, or is it necessary to explain and predict their occurrence? Is it possible or desirable to apply simple mathematical concepts to complicated ecological systems? Holling's most influential ideas have been presented in simple but vague formulations, and I can attest to the difficulty in deriving unambiguous tests of their validity or even in discussing these ideas with others who may see them from differing perspectives. The process of testing and rejecting or modifying and finally incorporating novel ideas into the mainstream of science reminds me of the old saying: "The mills of the gods grind slowly, but they grind exceedingly fine." We are somewhere in the middle of that process with regard to gaps and clumps.

Skillen and Maurer's chapter has solidified Allen, Forys, and Holling's (1999) results. Figure 11.4 is derived from the Breeding Bird Survey. It not only shows gap structure in body sizes of birds within six physiographic regions, but also indicates that many of these gaps occur at the same sizes within all or most of the regions. Skillen and Maurer also demonstrate that declining species are more likely to be near the edges of gaps and that nonindigenous species tend to be farther away from gaps. They point out: "With so much environmental heterogeneity present among these six ecosystems, it is difficult to envision a simple underlying mechanism. Yet the discontinuities in the assemblages are nonrandomly located, and the species react to these discontinuities in a nonrandom manner, which suggests that there must be some common underlying mechanism. We propose that the mechanism consists of the identifiable effects that the template of landscape structure has on population dynamics."

The explanation they offer is based largely on Holling's 1992 paper. I believe that Maurer is still concerned by the lack of details in Holling's assertions, but that he is willing to accept these assertions provisionally as a basis for further research. At the time of the Santa Fe meeting, Jim Brown also wrote to me:

> I remain skeptical about Buzz's [Holling's] ideas about "panarchy" and the "figure 8" [adaptive cycle] as a general model. I agree that some systems may (self-?)organize until they reach some critical state and then have a revolutionary, resetting collapse. But I think that this is far from universal. Many forest ecosystems with regular fire conform to this model, but I am not convinced that the deserts that I study or many other systems do this. I think that they continually reorganize in response to internal and external perturbations, but more or less gradually and without collapse or revolution. I am concerned that Buzz wants the community to buy into his complete worldview. That view is very creative, but I don't think that it will stand the test of time as an integrated whole. I also feel strongly that if he wants us to buy in and start pursuing these ideas, he has to make concepts like resilience, panarchy, innovation, etc., more precise and operational—i.e., expressed in mathematical terms and much more firmly rooted in the fundamentals of physics, chemistry, and biology.

On the basis of their remarks, I think that both Jim Brown and Brian Maurer can see great potential value in Holling's ideas, but they point out serious difficulties that will require enormous efforts to overcome. This volume, however, clearly demonstrates that there is something to be learned from gap and clump analysis. In chapter 1, Holling, Peterson, and Allen state, "Evidence is accumulating that body masses are distributed in a discontinuous manner both on land and in water, whose cause must be associated with slow, conservative properties of landscapes and waterscapes." This idea is labeled the Textural Discontinuity Hypothesis at several places in this volume, and several examples of evidence to support the hypothesis are offered as well.

In chapter 7, Restrepo and Arango examine range sizes instead of body sizes for North American birds and butterflies. Their main results appear in figures 7.4, 7.5, and 7.6. A cluster analysis reveals a small set of zooregions that have ecological significance. Remarkably, the distribution of range sizes shows lumps and gaps for both birds and butterflies, and important gaps are common to both. Restrepo and Arango state in their conclusions:

> The simultaneous examination of geographical range size and ecoregions provides tremendous insight into the processes underlying the distribution of attributes used to characterize species assemblages. In particular, the distribution of geographical range size has been characterized in most instances as a continuous, unimodal, right-skewed distribution that may become normal or slightly left skewed when log transformed. Instead, we found several aggregations of varying size clearly associated with landscape attributes. These findings may have implications in terms of

how we define endemic species, how we predict which geographical range sizes are likely to expand or contract, and perhaps which areas deserve special conservation status because many species seem to originate in them.

In chapter 5, Sendzimir offers another example of support for the Textural Discontinuity Hypothesis:

> If body-size distributions reflect mammal interactions with landscape structure, then certain body-size patterns would be consistently associated with particular landscape structures. I found that the location of aggregations and gaps of body-size distributions do tend to line up, but only when the size distributions come from species assemblages living in similar landscapes (i.e., within the same biome type). This consistency of body-size pattern correlation is greater than expected by chance . . . . Furthermore, two mammal assemblages with no species overlap that inhabit the same biome in Africa showed striking similarities in the locations of discontinuities and aggregations. In this case, it was not taxonomic similarity, but similarities in the scale of landscape architecture exploited that best predicted the locations of discontinuities and aggregations. Although evidence supports the TDH [Textural Discontinuity Hypothesis] at the macroscale of landscapes, it is quite reasonable to expect that multiple processes (trophic or phylogenetic or both) influence body-size patterns of assemblages that exploit landscape textures that range in scale from centimeters to kilometers.

He goes on to state:

> Discontinuities in the phylogenetically derived size template appear as animals interact with discontinuous, scale-specific landscape structure and either remain and reproduce or flee or die. I ascribe gaps in body-size patterns of mammal assemblages to discontinuities in the habitat architecture because landscape structure is the only variable that is consistently associated with similarities in the locations of aggregations and gaps. The discontinuities and aggregations in different body-size patterns tend to coincide if the two separate communities occur in very similar landscapes, and they do so far more frequently than expected by chance. Mammal body-size distributions enrich the picture of landscapes as complex adaptive systems by indexing the operation of two processes at different scales (phylogeny and animal interaction with habitat architecture) in the dynamic organization of species assemblages.

So far, statistical analysis of the Textural Discontinuity Hypothesis and similar hypotheses has been awkward because experiments are difficult or impossible. Hence, one must be content with "found data" and simply acknowledge that the prerequisites for many standard methods have not yet been met. In chapter 9, Stow, Sendzimir, and Holling offer a critique of various approaches and call for new methods to avoid the pitfall of "capricious" interpretation of results.

In chapter 6, Gunderson examines physical variables such as topography, temperature, and rainfall as well as vegetation variables in the Florida Everglades. He suggests that a small number of mixed physical and biological variables may be structuring ecosystems. If so, this process might lead to mechanistic explanations for patterns that appear in a variety of observable variables and hence might begin to address some of Maurer and Brown's reservations. Rosser's contribution, chapter 10, also leads to possible mechanistic explanations: he looks for dynamic instabilities in systems that are heavily influenced by human activities. In chapter 4, Marquet, Abades, Keymer, and Zeballos explore possible mechanisms involving the efficiency of energy acquisition and utilization, and they also point out the remarkable role of history in shaping present observations. This role is apparent in figure 4.7, where the collection of body masses for South American mammals is displayed together with the geographical origins of each species. Two peaks on the distribution of body masses coincide with corresponding peaks in the masses of North American mammals, and the intermediate peak comes from native South American fauna. The native South American species were isolated from the others before the development of the Isthmus of Panama 2.5 million years ago, so the present distribution reflects this history.

In chapter 3, Cumming and Havlicek also look for historical explanations: they show that "the existence of multiple modes in the distribution of body size and other characters can be explained as a simple consequence of the way that evolutionary processes operate and, in particular, by the two cornerstones of natural-selection theory, gradual descent and competition."

Another sort of progress is to extend the inquiry beyond body sizes or range sizes to other sorts of data. In chapter 8, Garmestani, Allen, and Bessey examine the structure of city sizes in the United States. They find evidence to support a "slaving principle" in which large-scale, slow processes (e.g., national economies) enslave small-scale, fast processes (e.g., regional and city economies). This principle is an analogue of Holling's Textural Discontinuity Hypothesis. Garmestani, Allen, and Bessey speculate that patterns of different nations' economic performance may also reflect underlying differences in social structure or possibly in history.

In light of these variations and analogues, the Textural Discontinuity Hypothesis now appears as a valuable diagnostic tool: patterns may be expected to appear in a variety of data related to size or area. The patterns themselves may help to classify systems, as when economists speak of "convergence clubs" to interpret clustering of gross domestic product data. Having found patterns, we then may look for mechanisms that may produce such patterns through entrainment with underlying structures in topography or climate or society or vegetation or interactions with inhabitants of the landscape. The mechanisms will undoubtedly involve dynamic phenomena such as instabilities and multiple stable states. This is not to say that patterns will be easy to detect or that that the underlying mechanisms will be apparent. On the contrary: we must always guard against "capricious" interpretations based on our built-in tendency to see pattern everywhere and to tell "just so" stories that will impede development of biological insight. This field is a mixture of flashes of insight and slow, never-ending attempts to winnow out structures and processes that underlie the patterns that appear. It is still controversial. I hope and expect, though, that it will always be a source of insight and challenge.

REFERENCES

Alig, R.J., D.M. Adams, and B.A. McCarl. 1998. Ecological and economic impacts of forest policies: Interactions across forestry and agriculture. *Ecological Economics* 27:63–78.

Allee, W.C. 1931. *Animal Aggregations.* Chicago: University of Chicago Press.

Allen, C.R. 2001. Ecosystems and immune systems: Hierarchical response provides resilience against invasions. *Conservation Ecology* 5 (1): 15. Available at: http://www.consecol.org/vol5/iss1/art15.

——. 2006. Predictors of introduction success in the South Florida avifauna. *Biological Invasions* 8:491–500.

Allen, C.R., E.A. Forys, and C.S. Holling. 1999. Body mass patterns predict invasions and extinctions in transforming landscapes. *Ecosystems* 2:114–121.

Allen, C.R., A. Garmestani, T. Havlicek, P. Marquet, G.D. Peterson, C. Restrepo, C. Stow, and B. Weeks. 2006. Patterns in body mass distributions: Sifting among alternative competing hypotheses. *Ecology Letters* 9:630–643.

Allen, C.R. and C.S. Holling. 2002. Cross-scale structure and scale breaks in ecosystems and other complex systems. *Ecosystems* 5:315–318.

Allen, C.R. and D.A. Saunders. 2002. Variability between scales: Predictors of nomadism in birds of an Australian Mediterranean-climate ecosystem. *Ecosystems* 5:348–359.

——. 2006. Multimodel inference and the understanding of complexity, discontinuity, and nomadism. *Ecosystems* 9:694–699.

Allen, T.F.H. and T.W. Hoekstra. 1992. *Toward a Unified Ecology.* New York: Columbia University Press.

Allen, T. F .H. and T.B. Starr. 1982. *Hierarchy: Perspectives for Ecological Complexity.* Chicago: University of Chicago Press.

Alroy, J. 1998. Cope's rule and the dynamics of body mass evolution in North American fossil mammals. *Science* 280:731–734.

Anderson, L.G. 1973. Optimum economic yield of a fishery given a variable price of output. *Journal of the Fisheries Research Board of Canada* 30:509–513.

Angermeir, P.L. 1995. Ecological attributes of extinction-prone species: Loss of freshwater fishes of Virginia. *Conservation Biology* 9:143–158.

Aoki, M. 1996. *New Approaches to Macroeconomic Modelling: Evolutionary Stochastic Dynamics, Multiple Equilibria, and Externalities as Field Effects.* Cambridge, U.K.: Cambridge University Press.

Arita, H.T., F. Figueroa, A. Frisch, P. Rodriguez, and K. Santos Del Prado. 1997. Geographical range size and the conservation of Mexican mammals. *Conservation Biology* 11:92–100.

Arrow, K.J. and A. C. Fisher. 1974. Preservation uncertainty and irreversibility. *Quarterly Journal of Economics* 87:312–319.

August, P., L. Iverson, and J. Nugranad. 2002. Human conversion of terrestrial habitats. Pages 198–224 in K. J. Gutzwiller, ed., *Applying Landscape Ecology in Biological Conservation*. New York: Springer.

Bailey, R.G. 1998. *Ecoregions Map of North America*. U.S. Department of Agriculture (USDA) Forest Service Miscellaneous Publication no.1548. Washington, D.C.: USDA.

Bak, P. 1996. *How Nature Works: The Science of Self-Organized Criticality*. New York: Copernicus.

Bak, P., C. Tang, and K. Wiesenfeld. 1988. Self-organized criticality. *Physical Review* A 38:364–374.

Baker, W.L. 1993. Spatially heterogeneous multiscale response of landscapes to fire suppression. *Oikos* 66:66–71.

——. 1995. Long-term response of disturbance landscapes to human intervention and global change. *Landscape Ecology* 10:143–159.

Barnosky, A.D., P.L. Koch, R.S. Feranec, S.L. Wing, and A.B. Shabel. 2004. Assessing the causes of late Pleistocene extinctions on the continents. *Science* 306:70–75.

Barro, R.J. 1997. *Determinants of Economic Growth: A Cross-Country Empirical Study*. Cambridge, Mass.: MIT Press.

Barro, R.J. and X. Salai-i-Martin. 1992. Convergence. *Journal of Political Economy* 100:223–251.

Barry, R.G. and R. Chorley. 1992. *Atmosphere, Weather, and Climate*. 6th ed. New York: Routledge.

Barton, K.E., D.G. Howell, and J.F. Vigil. 2003. *The North America Tapestry of Time and Terrain*. U.S. Geological Survey (USGS) Geological Investigation Series I-2781. Washington, D.C.: USGS.

Bascompte, J. and R. Solé. 1996. Habitat fragmentation and extinction thresholds in spatially explicit models. *Journal of American Ecology* 65:465–473.

Begon, M., J.L. Harper, and C.R. Townsend. 1990. *Ecology: Individuals, Populations, and Communities*. 2d ed. London: Blackwell Scientific.

Bellwood, D.R., T.P. Hughes, C. Folke, and M. Nystrom. 2004. Confronting the coral reef crisis. *Nature* 429:827–833.

Belovski, G.E., D.B. Botkin, T.A. Crowl, K.W. Cummins, J.F. Franklin, M.L. Hunter Jr., A. Joern, D.B. Lindenmayer, J.A. MacMahon, C.R. Margules, and J.M. Scott. 2004. Ten suggestions to strengthen the science of ecology. *BioScience* 54:345–351.

Bennett, E.M., S.R. Carpenter, and N.F. Caraco. 2001. Human impacts on erodable phosphorus and eutrophication: A global perspective. *BioScience* 51:227–234.

Berkes, F. and C. Folke. 1998. *Linking Social and Ecological Systems*. Cambridge, U.K.: Cambridge University Press.

Berry, B.J.L. 1971. City size and economic development: Conceptual synthesis and policy problems, with special reference to South and Southeast Asia. Pages 111–155 in L. Jakobson and V. Prakash, eds., *Urbanization and National Development*. Beverly Hills, Calif.: Sage.

Bessey, K.M. 2000. Scale, structure, and dynamics in the U.S. urban systems, 1850–1990: City size in the lens of region. Ph.D. diss., Harvard University.

——. 2002. Structure and dynamics in an urban landscape: Toward a multiscale view. *Ecosystems* 5:360–375.

Bischi, G-I. and M. Kopel. 2002. The role of competition, expectations, and harvesting costs in commercial fishing. Pages 85–109 in T. Puu and I. Sushko, eds., *Oligopoly Dynamics: Model and Tools*. Heidelberg, Germany: Springer-Verlag.

Blackburn, T.M. and K.J. Gaston. 1994a. Animal body-size distributions—patterns, mechanisms, and implications. *Trends in Ecology and Evolution* 9:471–474.

——. 1994b. The distribution of body sizes of the world's bird species. *Oikos* 70:127–130.

Bokma, F. 2001. Evolution of body size: Limitations of an energetic definition of fitness. *Functional Ecology* 15:696–699.

——. 2002. A statistical test of unbiased evolution of body size in birds. *Evolution* 56:2499–2504.

Bonabeau, E. 1998. Social insect colonies as complex adaptive systems. *Ecosystems* 1:437–443.

Bond, W.J. and B.W. van Wilgen. 1996. *Fire and Plants*. London: Chapman and Hall.

Bradbury, R.H. and R.E. Reichelt. 1983. Fractal dimension of a coral reef at ecological scales. *Marine Ecology* 10:169–171.

Bradbury, R.H., R.E. Reichelt, and D.G. Green. 1984. Fractals in ecology: Methods and interpretation. *Marine Ecology* 14:295–296.

Bradley, J.T. 1972. *Climate of Florida*. Silver Springs, Md.: Climate of the States, Environmental Data Service.

Brauer, F. and A.E. Soudack. 1979. Stability regions and transition phenomena for harvesting predator-prey systems. *Journal of Mathematical Biology* 7:319–337.

——. 1985. Optimal harvesting in predator-prey systems. *International Journal of Control* 41:111–128.

Breiman, L., J.H. Friedman, R.A. Olshen, and C.J. Stone. 1984. *Classification and Regression Trees*. Belmont, Calif.: Wadsworth International Group.

Brewer, A.M. and K.J. Gaston. 2003. The geographical range structure of the holly leafminer. II. Demographic rates. *Journal of Animal Ecology* 72:82–93.

Brock, W.A. 1993. Pathways to randomness in the economy: Emergent nonlinearity in economics and finance. *Estudios Económicos* 8:3–55.

——. 1999. Scaling in economics: A reader's guide. *Industrial and Corporate Change* 8:409–446.

Brock, W. and D. Evans. 1986. *The Economics of Small Businesses: Their Role and Regulation in the U.S. Economy*. New York: Holmes and Meier.

Bromley, D.W. 1991. *Environment and Economy: Property Rights and Public Policy*. Oxford, U.K.: Basil Blackwell.

Brooks, M.L., C.M. D'Antonio, D.M. Richardson, J.B. Grace, J.E. Keeley, J.M. DiTomaso, R.J. Hobbs, M. Pellant, and D. Pyke. 2004. Effects of invasive alien plants on fire regimes. *BioScience* 54:677–688.

Brown, J.H. 1984. On the relationship between distribution and abundance. *American Naturalist* 124:255–279.

——. 1995. *Macroecology*. Chicago: University of Chicago Press.

——. 1999. Macroecology: Progress and prospect. *Oikos* 87:3–14.

Brown, J.H., J.F. Gillooly, A.P. Allen, V.M. Savage, and G.B. West. 2004. Toward a metabolic theory of ecology. *Ecology* 85:1771–1789.

Brown, J.H., P.A. Marquet, and M.L. Taper. 1993. Evolution of body size: Consequences of an energetic definition of fitness. *American Naturalist* 142:573–584.

Brown, J. H. and B. A. Maurer. 1987. Evolution of species assemblages: Effects of energetic constraints and species dynamics on the diversification of the North American avifauna. *American Naturalist* 130:1–17.

——. 1989. Macroecology: The division of food and space among species on continents. *Science* 243:1145–1150.

Brown, J. H. P. and P. F. Nicoletto. 1991. Spatial scaling of species composition: Body masses of North American land mammals. *American Naturalist* 138:1478–1512.

Brown, J. H., G. C. Stevens, and D. M. Kaufman. 1996. The geographic range: Size, shape, boundaries, and internal structure. *Annual Review of Ecology and Systematics* 27:597–623.

Brown, J. H., M. L. Taper, and P. A. Marquet. 1996. Darwinian fitness and reproductive power: Reply. *American Naturalist* 147:1092–1097.

Brundsen, D. 1996. Geomorphological events and landform change. *Zeitschrift für Geomorphologie* N. F. 40:273–288.

Bystrak, D. 1981. The North American Breeding Bird Survey. *Studies in Avian Biology* 6:34–41.

Calder, W. A. 1996. *Size, Function, and Life-History.* Mineola, N.Y.: Courier Dover.

Carbone, C. and J. L. Gittleman. 2002. A common rule for the scaling of carnivore density. *Science* 295:2273–2276.

Carbone, C., G. M. Mace, S. C. Roberts, and D. W. MacDonald. 1999. Energetic constraints on the diet of terrestrial carnivores. *Nature* 402:286–288.

Carpenter, S. R. 2000. Alternate states of ecosystems: Evidence and its implications for environmental decisions. Pages 357–383 in M. C. Press, N. Huntley, and S. Levin, eds., *Ecology: Achievement and Challenge.* London: Blackwell.

Carpenter, S. R. and W. A. Brock. 2004. Spatial complexity, resilience, and policy diversity: Fishing on lake-rich landscapes. *Ecology and Society* 9 (1): 8. Available at: http://www.ecologyandsociety.org/vol9/iss1/art8.

Carpenter, S. R., W. A. Brock, and D. Ludwig. 2002. Collapse, learning, and renewal. Pages 173–193 in L. H. Gunderson and C. S. Holling, eds., *Panarchy: Understanding Transformations in Human and Natural Systems.* Washington, D.C.: Island Press.

Carpenter, S. R., N. F. Caraco, D. L. Correll, R. W. Howarth, A. N. Sharpley, and V. H. Smith. 1998. Nonpoint pollution of surface waters with phosphorus and nitrogen. *Ecological Applications* 8:559–568.

Carpenter, S. R. and J. F. Kitchell. 1993. *The Trophic Cascade in Lakes.* Cambridge, U.K.: Cambridge University Press.

Carpenter, S. R., D. Ludwig, and W. A. Brock. 1999. Management of eutrophication for lakes subject to potentially irreversible change. *Ecological Applications* 9:751–771.

Caughley, G., D. Grice, R. J. Barker, and B. Brown. 1988. The edge of the range. *Journal of Animal Ecology* 57:771–785.

Chambers, J. M., W. S. Cleveland, K. Beat, and P. A. Tukey. 1983. *Graphical Methods for Data.* Boston: Duxbury Press.

Chapin, F. S., III, B. H. Walker, R. J. Hobbs, D. U. Hooper, J. H. Lawton, O. E. Sala, and D. Tilman. 1997. Biotic control over the functioning of ecosystems. *Science* 277:500–504.

Charles, A. T. 1988. Fishery socioeconomics: A survey. *Land Economics* 64:276–295.

Chipman, H. A, E. I. George, and R. E. McCulloch. 1998. Bayesian CART model search. *Journal of the American Statistical Association* 93:935–948.

Chown, S. L. and K. J. Gaston. 1997. The species–body size distribution: Energy, fitness, and optimality. *Functional Ecology* 11:365–375.

Ciriacy-Wantrup, S. V. and R. C. Bishop. 1975. "Common property" as a concept in natural resources policy. *Natural Resources Journal* 15:713–727.

Clark, C. W. 1973. Profit maximization and the extinction of animal species. *Journal of Political Economy* 81:363–372.

——. 1985. *Bioeconomic Modelling and Fisheries Management.* New York: Wiley-Interscience.

——. 1990. *Mathematical Bioeconomics: The Optimal Management of Renewable Resources.* 2d ed. New York: Wiley-Interscience.

Clark, C. W., F. H. Clarke, and G. R. Munro. 1979. The optimal exploitation of renewable resource stocks: Problems of irreversible investment. *Econometrica* 47:25–49.

Clark, C. W. and M. Mangel. 1979. Aggregation and fishery dynamics: A theoretical study of schooling and the purse seine tuna fisheries. *Fisheries Bulletin* 77:317–337.

——. 2000. *Dynamic State Variables in Ecology: Methods and Applications.* Oxford, U.K.: Oxford University Press.

Clark, J. S. and D. Pregibon. 1992. Tree-based models. Pages 377–419 in J. M. Chambers and T. J. Hastie, eds., *Statistical models in S.* Pacific Grove, Calif.: Wadsworth and Brooks/Cole.

Clark, W. C. 1985. Scales of climate impacts. *Climate Change* 7:5–27.

Clark, W. C., D. D. Jones, and C. S. Holling. 1979. Lessons for ecological policy design: A case study of ecosystem management. *Ecological Modelling* 7:2–53.

Cohen, J. E., T. Jonsson, and S. R. Carpenter. 2003. Ecological community description using the food web, species abundance, and body size. *Proceedings of the National Academy of Science* 100:1781–1786.

Cohen, J. E., S. L. Pimm, P. Yodzis, and J. Saldana. 1993. Body sizes of animal predators and animal prey in food webs. *Journal of Animal Ecology* 62:67–78.

Conklin, J. E. and W. C. Kolberg. 1994. Chaos for the halibut. *Marine Resource Economics* 9:159–182.

Connell, J. H. and W. P. Sousa. 1983. On the evidence needed to judge ecological stability or persistence. *American Naturalist* 121:789–824.

Copes, P. 1970. The backward-bending supply curve of the fishing industry. *Scottish Journal of Political Economy* 17:69–77.

Cornelius, J. M. and J. F. Reynolds. 1991. On determining the statistical significance of discontinuities within ordered ecological data. *Ecology* 72:2057–2070.

Cox, P. M., R. A. Betts, C. D. Jones, S. A. Spall, and I. J. Totterdell. 2000. Acceleration of global warming due to carbon-cycle feedbacks in a coupled climate model. *Nature* 408:184–187.

Crisp, M. D., S. Laffan, H. P. Linder, and A. Monro. 2001. Endemism in the Australian flora. *Journal of Biogeography* 28:183–198.

Crooks, K. R. and M. E. Soule. 1999. Mesopredator release and avifaunal extinctions in a fragmented system. *Nature* 400:563–566.

Cuddington, K. and P. Yodzis. 2002. Predator-prey dynamics and movement in fractal environments. *American Naturalist* 160:119–134.

Cumming, G. S. and T. Havlicek. 2002. Evolution, ecology, and multimodal distributions of body size. *Ecosystems* 5:705–711.

Curnutt, J. L., S. L. Pimm, and B. A. Maurer. 1996. Population variability of sparrows in space and time. *Oikos* 76:131–144.

Cyr, H., J. A. Downing, and R. H. Peters. 1997. Density body size relationships in local aquatic communities. *Oikos* 79:333–340.

Cyr, H. and R. H. Peters. 1996. Biomass-size spectra and the prediction of fish biomass in lakes. *Canadian Journal of Fisheries and Aquatic Sciences* 53:994–1006.

Damuth, J. 1981. Population density and body size in mammals. *Nature* 190:699–700.

——. 1987. Interspecific allometry of population density in mammals and other animals: The independence of body mass and population energy use. *Biological Journal of the Linnean Society* 31:193–246.

D'Antonio, C. M. and P. M. Vitousek. 1992. Biological invasions by exotic grasses, the grass-fire cycle, and global change. *Annual Review of Ecology and Systematics* 3:63–87.

Darwin, C. 1859. *On the Origin of Species by Means of Natural Selection or the Preservation of Favoured Races in the Struggle for Life.* London: John Murray.

Davis, S. M., L. H. Gunderson, W. Park, J. Richardson, and J. Mattson. 1994. Landscape dimension, composition, and function in a changing Everglades ecosystem. Pages 419–444 in S. M. Davis and J. C. Ogden, eds., *The Everglades: Spatial and Temporal Patterns for Ecosystem Restoration.* St Lucie, Fla.: St. Lucie Press.

Dawkins, R. 1996. *The Blind Watchmaker.* New York: W. W. Norton.

de Klerk, H. M., T. M. Crowe, J. Fjeldsa, and N. D. Burgess. 2002. Patterns of species richness and narrow endemism of terrestrial bird species in the Afrotropical region. *Journal of Zoology* 256:327–342.

Delcourt, H. R. 1987. The impact of prehistoric agriculture and land occupation on natural vegetation. *Trends in Ecology and Evolution* 2:39–44.

Demment, M. W. and P. J. Van Soest. 1985. A nutritional explanation for body-size patterns of rumminant and nonrumminant herbivores. *American Naturalist* 125:641–672.

Dendrinos, D. S. 1992. *The Dynamics of Cities: Ecological Determinism, Dualism, and Chaos.* London: Routledge.

Dendrinos, D. S. and M. Sonis. 1990. *Chaos and Socio-spatial Dynamics.* New York: Springer.

Dennis, B. 1989. Stochastic differential equations as insect population models. Pages 219–238 in L. McDonald, B. Manly, J. Lockwood, and J. Logan, eds., *Estimation and Analysis of Insect Populations.* Berlin: Springer-Verlag.

Dennis, B. and R. F. Costantino. 1988. Analysis of steady-state populations with the gamma-abundance model: Application to Tribolium. *Ecology* 69:1200–1213.

Dennis, B. and G. P. Patil. 1984. The gamma-distribution and weighted multimodal gamma-distributions as models of population abundance. *Mathematical Biosciences* 68:187–212.

Diana, J. S. 1995. *Biology and Ecology of Fishes.* Traverse City, Mich.: Cooper.

Diener, M. and T. Poston. 1984. The perfect delay convention, or the revolt of the slaved variables. Pages 249–268 in H. Haken, ed., *Chaos and Order in Nature,* 2d ed. Berlin: Springer-Verlag.

Done, T. J. 1992. Phase shifts in coral reef communities and their ecological significance. *Hydrobiology* 247:121–132.

Dow, K. 2000. Social dimensions of gradients in urban ecosystems. *Urban Ecosystems* 4:255–275.

Dublin, H. T., A. R. E. Sinclair, and J. McGlade. 1990. Elephants and fire as causes of multiple stable states in the Serengeti-Mara woodlands. *Journal of Animal Ecology* 59:1147–1164.

Dubost, G. 1979. The size of the African forest artiodactyls as determined by vegetation structure. *African Journal of Ecology* 17:1–17.

Duncan, R.P., T.M. Blackburn, and C.J. Veltman. 1999. Determinants of geographical range sizes: A test using introduced New Zealand birds. *Journal of Animal Ecology* 68:963–975.

Dunning, J.B., Jr., ed. 1993. *CRC Handbook of Avian Body Masses.* Boca Raton, Fla.: CRC Press.

Durlauf, S.N. and D. Quah. 1999. The new empirics of economic growth. Pages 235–308 in J. Taylor and M. Woodford, eds., *Handbook of Macroeconomics.* Amsterdam: North Holland.

Durrenberger, E.P. and G. Pàlsson. 1987. Resource management in Icelandic fishing. Pages 370–392 in B.J. McCay and J.M. Acheson, eds., *The Question of the Commons.* Tucson: University of Arizona Press.

Dziewonski, K. 1972. General theory of rank-size distributions in regional settlement systems: Reappraisal and reformulation of the rank-size rule. *Papers of the Regional Science Association* 29:73–86.

Efron, B. and R.J. Tibshirani. 1993. *An Introduction to the Bootstrap.* New York: Chapman and Hall.

Ehrlich, P.R., D.S. Dobkin, and D. Wheye. 1988. *The Birder's Handbook: A Field Guide to the Natural History of North American Birds.* New York: Simon and Schuster.

Elliott, J.J. and R.S. Arbib Jr. 1953. Origin and status of the House Finch in the eastern United States. *The Auk* 70:31–37.

Elmqvist, T., C. Folke, M. Nystrom, G. Peterson, J. Bengtsson, B. Walker, and J. Norberg. 2003. Response diversity, ecosystem change, and resilience. *Frontiers in Ecology and the Environment* 1:488–494.

Elton, C.S. 1958. *The Ecology of Invasions by Plants and Animals.* London: Methuen.

Enquist, B.J., E.P. Economo, T.E. Huxman, A.P. Allen, D.D. Ignace, and J.F. Gillooly. 2003. Scaling metabolism from organisms to ecosystems. *Nature* 423:639–642.

Estes, J.A. and D. Duggins. 1995. Sea otters and kelp forests in Alaska: Generality and variation in a community ecological paradigm. *Ecological Monographs* 65:75–100.

Faustmann, M. [1849] 1968. On the determination of the value which forest land and immature stands possess. English translation in M. Gane, ed., *Martin Faustmann and the Evaluation of Discounted Cash Flow.* Oxford University Paper no. 42. Oxford, U.K.: Oxford University.

Feder, J. 1988. *Fractals.* New York: Plenum Press.

Fenneman, N.M. and D.W. Johnson. 1946. *Physical Divisions of the United States.* U.S. Geological Survey (USGS) Physiography Committee Special Map, Scale 1:7,000,000. Washington, D.C.: USGS

Fishbase: A Global Information System on Fishes. 2000. Multiauthor on-line database available at: http://www.fishbase.org/home.htm.

Fitzgerald, J.A., J.R. Herkert, and J.D. Brawn. 2000. *Partners in Flight Bird Conservation Plan for the Prairie Peninsula (Physiographic Area 31).* Brentwood, Mo.: Partners in Flight.

Flannery, T.F. 2002. *The Eternal Frontier.* New York: Grove Press.

Foley, J.A., M.H. Costa, C. Delire, N. Ramankutty, and P. Snyder. 2003. Green surprise? How terrestrial ecosystems could affect earth's climate. *Frontiers in Ecology and the Environment* 1:38–44.

Folke, C., S. Carpenter, T. Elmqvist, L. Gunderson, C.S. Holling, and B. Walker. 2002. Resilience and sustainable development: Building adaptive capacity in a world of transformations. *Ambio* 31:437–440.

Forys, E. A. and C. R. Allen. 1999. Biological invasions and deletions: Community change in South Florida. *Biological Conservation* 87:341–347.

———. 2002. Functional group change within and across scales following invasions and extinctions in the Everglades ecosystem. *Ecosystems* 5:339–347.

Friedberg, S. H. and A. J. Insel. 1986. *Introduction to Linear Algebra with Applications.* Englewood Cliffs, N.J.: Prentice-Hall.

Fuller, W. A. 1987. *Measurement Error Models.* New York: John Wiley and Sons.

Gabaix, X. 1999. Zipf's law for cities: An explanation. *Quantitative Journal of Economics* 3:739–767.

Gannon, P. T. 1978. *Influence of Earth Surface and Cloud Properties on South Florida Sea Breeze.* National Oceanic and Atmospheric Administration (NOAA) Technical Report ERL 402-NHELM2. Washington, D.C.: NOAA.

Gardezi, T. and J. da Silva. 1999. Diversity in relation to body size in mammals: A comparative study. *American Naturalist* 153:110–123.

Garmestani, A. S., C. R. Allen, and K. M. Bessey. 2005. Time series analysis of clusters in city size distributions. *Urban Studies* 42:1507–1515.

Garmestani, A. S., C. R. Allen, J. D. Mittelstaedt, C. A. Stow, and W. A. Ward. 2006. Firm size diversity, functional richness, and resilience. *Environment and Development Economics* 11:533–551.

Gaston, K. J. 1990. Patterns in the geographical ranges of species. *Biological Reviews of the Cambridge Philosophical Society* 65:105–129.

———. 1994. Measuring geographic range sizes. *Ecography* 17:198–205.

———. 1996. Species-range-size distributions: Patterns, mechanisms, and implications. *Trends in Ecology and Evolution* 11:197–201.

———. 1998. Species–range size distributions: Products of speciation, extinction, and transformation. *Philosophical Transactions of the Royal Society of London* Series B: Biological Sciences 353:219–230.

Gaston, K. J. and T. M. Blackburn. 1997. Age, area, and avian diversification. *Biological Journal of the Linnean Society* 62:239–253.

———. 2000. *Pattern and Process in Macroecology.* Oxford, U.K.: Blackwell Science.

Gaston, K. J. and F. L. He. 2002. The distribution of species range size: A stochastic process. *Proceedings of the Royal Society of London* Series B: Biological Sciences 269: 1079–1086.

Gaston, K. J., R. M. Quinn, T. M. Blackburn, and B. C. Eversham. 1998. Species–range size distributions in Britain. *Ecography* 21:361–370.

Gaston, K. J., R. M. Quinn, S. Wood, and H. R. Arnold. 1996. Measures of geographic range size: The effects of sample size. *Ecography* 19:259–268.

Gause, G. F. 1935. Experimental demonstration of Volterra's periodic oscillations in the numbers of animals. *Journal of Experimental Biology* 12:44–48.

Gell-Mann, M. 1994. *The Quark and the Jaguar.* New York: Freeman.

Gentry, R. C. 1984. Hurricanes in south Florida. Pages 510–517 in P. J. Gleason, ed., *Environments of South Florida: Present and Past II, Memoir 2.* Coral Gables, Fla.: Miami Geological Society.

Gibrat, R. 1957. On economic inequalities. *International Economic Papers* 7:53–70.

Ginzburg, L. and M. Colyvan. 2004. *Ecological Orbits: How Planets Move and Populations Grow.* Oxford, U.K.: Oxford University Press.

Gleason, P. J. 1972. The origin, sedimentation, and stratigraphy of a calcitic mud located in the southern freshwater Everglades. Ph.D. diss., Pennsylvania State University.

Gleason, P.J., A.D. Cohen, P. Stone, W.G. Smith, H.K. Brooks, R. Goodrick, and W. Spackman Jr. 1984. The environmental significance of Holocene sediments from the Everglades and saline tidal plains. Pages 287–341 in P.J. Gleason, ed., *Environments of South Florida: Present and Past II, Memoir 2.* Coral Gables, Fla.:Miami Geological Society.

Glitzenstein, J.S., W.J. Platt, and D.R. Streng. 1995. Effects of fire regime and habitat on tree dynamics in north Florida longleaf pine savannas. *Ecological Monographs* 65:441–476.

Gordon, H.S. 1954. Economic theory of a common property resource: The fishery. *Journal of Political Economy* 62:124–142.

Goss-Custard, J.D., R.M. Warwick, R. Kirby, S. McGrorty, R.T. Clarke, B. Pearson, W.E. Rispin, S.E.A. Le V. Dit Durell, and R.J. Rose. 1991. Towards predicting wading bird densities from predicted prey densities in a post-barrage severn estuary. *Journal of Applied Ecology* 28:1004–1026.

Gould, S.J. 2002. *The Structure of Evolutionary Theory.* Cambridge, Mass.: Belknap Press of Harvard University Press.

Grebogi, C., E. Ott, and J.A. Yorke. 1987. Chaos, strange attractors, and fractal basin boundaries in nonlinear dynamics. *Science* 238:632–638.

Greenwood, P. H. 1981. Species flocks and explosive evolution. Pages 64–74 in P.H. Greenwood and P.L. Forey, eds., *Chance, Change, and Challenge—the Evolving Biosphere.* London: Cambridge University Press and British Museum of Natural History.

Grimm, N.B., J.M. Grove, S.T.A. Pickett, and C.L. Redman. 2000. Integrated approaches to long-term studies of urban ecological systems. *BioScience* 50: 571–593.

Gunderson, L.H. 1992. Spatial and temporal hierarchies in the Everglades ecosystem with implications for water management. Ph.D. diss., University of Florida, Gainesville.

——. 1994. Vegetation of the Everglades. Pages 323–340 in S.M. Davis and J.C. Ogden, eds., *The Everglades: Spatial and Temporal Patterns for Ecosystem Restoration.* St. Lucie, Fla.: St. Lucie Press.

——. 1999. Resilience, flexibility, and adaptive management—antidotes for spurious certitude? *Conservation Ecology* 3 (1): 7. Available at: http://www.consecol.org/vol3/iss1/art7/.

Gunderson, L.H. and C. S. Holling. 2002. *Panarchy: Understanding Transformations in Human and Natural Systems.* Washington, D.C.: Island Press.

Gunderson, L.H., C.S. Holling, L. Pritchard Jr., and G. Peterson. 2002. Understanding resilience: Theory, metaphors, and frameworks. Pages 3–20 in L.H. Gunderson and L. Pritchard Jr., eds., *Resilience and the Behavior of Large-Scale Systems.* Washington, D.C.: Island Press.

Gunderson, L.H. and W.F. Loftus. 1993. The Everglades: Competing land uses imperil the biotic communities of a vast wetland. Pages 199–255 in S.C. Boyce, W.H. Martin, and A.C. Echternacht, eds., *Biotic Diversity of the Southeastern United States.* New York: John Wiley and Sons.

Gunderson, L.H. and L. Pritchard. 2002. *Resilience and the Behavior of Large Scale Ecosystems.* Washington, D.C.: Island Press.

Hagmeier, E.M. and C.D. Sults. 1964. A numerical analysis of the distributional patterns of North American mammals. *Systematic Zoology* 13:125–155.

Hall, P. and M. York. 2001. On the calibration of Silverman's test for multimodality. *Statistica Sinica* 11:515–536.

Hanski, I. 1982. Dynamics of regional distribution: The core and satellite species hypothesis. *Oikos* 38:210–211.

Harestad, A. S. and F. Bunnel. 1979. Home range and body weight—a reevaluation. *Ecology* 60:389–402.

Harris, G. P., B. B. Piccinin, and J. van Ryn. 1983. Physical variability and phytoplankton communities: V. Cell size, niche diversification, and the role of competition. *Archive für Hydrobiologie* 98:215–239.

Hartman, R. 1976. The harvesting decision when a standing forest has value. *Economic Inquiry* 14:52–58.

Hartvigsen, G., A. Kinzig, and G. Peterson. 1998. Use and analysis of complex adaptive systems in ecosystem science: Overview of special section. *Ecosystems* 1:427–430.

Havlicek, T. D. and S. R. Carpenter. 2001. Pelagic species size distributions in lakes: Are they discontinuous? *Limnology and Oceanography* 46:1021–1033.

Hela, I. 1952. Remarks on the climate of South Florida. *Bulletin of Marine Science* 2:438–447.

Henderson, J. V. 1974. The sizes and types of cities. *American Economic Review* 64: 640–656.

Heyward, F. 1939. The relation of fire to stand composition of longleaf pine forests. *Ecology* 20:287–304.

Hilborn, R. and M. Mangel. 1997. *The Ecological Detective: Confronting Models with Data.* Princeton, N.J.: Princeton University Press.

Hoffmeister, J. E. 1974. *Land from the Sea.* Coral Gables, Fla.: University of Miami Press.

Holland, J. H. 1995. *Hidden Order: How Adaptation Builds Complexity.* Reading, Mass.: Addison Wesley.

Holland, J. H., K. J. Holyoak, R. E. Nisbett, and P. R. Thagard. 1989. *Induction: Processes of Inference, Learning, and Discovery.* Cambridge, Mass.: MIT Press.

Holling, C. S. 1965. The functional response of predators to prey density and its role in mimicry and population regulation. *Memoirs of the Entomological Society of Canada* 45:1–60.

——. 1973. Resilience and stability of ecological systems. *Annual Review of Ecology and Systematics* 4:1–24.

——. 1986. Resilience of ecosystems: Local surprise and global change. Pages 292–317 in W. C. Clark and R. E. Munn, eds., *Sustainable Development of the Biosphere.* Cambridge, U.K.: Cambridge University Press.

——. 1988. Temperate forest insect outbreaks, tropical deforestation, and migratory birds. *Memoirs of the Entomological Society of Canada* 146:22–32.

——. 1992. Cross-scale morphology, geometry, and dynamics of ecosystems. *Ecological Monographs* 62:447–502.

——. 2001. Understanding the complexity of economic, ecological, and social systems. *Ecosystems* 4:390–405.

Holling, C. S. and C. R. Allen. 2002. Adaptive inference for distinguishing credible from incredible patterns in nature. *Ecosystems* 5:319–328.

Holling, C. S. and L. H. Gunderson. 2002. Resilience and adaptive cycles. Pages 25–62 in L. H. Gunderson and C. S. Holling, eds., *Panarchy: Understanding Transformations in Human and Natural Systems.* Washington, D.C.: Island Press.

Holling, C. S., G. Peterson, P. Marples, J. Sendzimir, K. H. Redford, L. Gunderson, and D. Lambert. 1996. Self-organization in ecosystems: Lumpy geometries, periodicities, and morphologies. Pages 346–384 in B. Walker and W. Steffen, eds., *Global Change and Terrestrial Ecosystems*. Cambridge, U.K.: Cambridge University Press.

Hommes, C.H. and J.B. Rosser Jr. 2001. Consistent expectations equilibria and complex dynamics in renewable resource markets. *Macroeconomic Dynamics* 5: 180–203.

Hughes, T.P. 1994. Catastrophes, phase shifts, and large-scale degradation of a Caribbean coral reef. *Science* 265:1547–1551.

Hughes, T.P., D.R. Bellwood, and S.R. Connolly. 2002. Biodiversity hotspots, centres of endemicity, and the conservation of coral reefs. *Ecology Letters* 5:775–784.

Hunter, W.C., L.H. Peoples, and J.A. Collazo. 2001. *South Atlantic Coastal Plain Partners in Flight Bird Conservation Plan (Physiographic Area #03)*. Atlanta, Ga.: U.S. Fish and Wildlife Service in cooperation with Partners in Flight.

Huston, M.A. 1994. *Biological Diversity: The Coexistence of Species on Changing Landscapes*. Cambridge, U.K.: Cambridge University Press.

Hutchinson, G.E. 1948. Circular causal systems in ecology. *Annals of the New York Academy of Sciences* 50:221–246.

——. 1957. Concluding remarks. *Cold Spring Harbor Symposia on Quantitative Biology* 22:415–427.

——. 1959. Homage to Santa Rosalia, or Why are there so many kinds of animals? *American Naturalist* 93:145–159.

Hutchinson, G.E. and R.H. MacArthur. 1959. A theoretical ecological model of size distributions among species of animals. *American Naturalist* 93:117–125.

Isaacs, J.A. 1980. Precipitation regimes of Florida: Spatial analyses and time series. Master's thesis, University of Florida, Gainesville.

Jablonski, D. 1996. Body size and macroevolution. Pages 256–289 in D. Jablonski, H. Erwin, and J. H. Lipps, eds., *Evolutionary Paleobiology*. Chicago: University of Chicago Press.

Jablonski, D. and K. Roy. 2003. Geographical range and speciation in fossil and living molluscs. *Proceedings of the Royal Society of London* Series B: Biological Sciences 270:401–406.

Jackson, J.B.C., M.X. Kirby, W.H. Berger, K.A. Bjorndal, L.W. Botsford, B.J. Bourque, R.H. Bradbury, R. Cooke, J. Erlandson, J.A. Estes, T.P. Hughes, S. Kidwell, C.B. Lange, H.S. Lenihan, J.M. Pandolfi, C.H. Peterson, R.S. Steneck, M.J. Tegner, and R.R. Warner. 2001. Historical overfishing and the recent collapse of coastal ecosystems. *Science* 293:629–637.

Jacobs, B.F. 1999. Estimation of rainfall variables from leaf characteristics in tropical Africa. *Palaeogeography, Palaeoclimatology, Palaeoecology* 145:231–250.

Jain, S. and S. Krishna. 2002. Large extinctions in an evolutionary model: The role of innovation and keystone species. *Proceedings of the National Academy of Science, USA* 99:2055–2060.

Jaksic, F.M., P. Feinsinger, and J. E. Jimenez. 1996. Ecological redundancy and long-term dynamics of vertebrate predators in semiarid Chile. *Conservation Biology* 10: 252–262.

Jansson, R. 2003. Global patterns in endemism explained by past climatic change. *Proceedings of the Royal Society of London* Series B: Biological Sciences 270:583–590.

Jefferson, M. 1939. The law of the primate city. *Geographical Review* 29:226–232.

Johnson, K.N., D.B. Jones, and B.M. Kent. 1980. *A User's Guide to the Forest Planning Model (FORPLAN)*. Fort Collins, Colo.: U.S. Department of Agriculture (USDA) Forest Service, Land Management Planning.

Jones, D. D. and C. J. Walters. 1976. Catastrophe theory and fisheries regulation. *Journal of the Fisheries Resource Board of Canada* 33:2829–2833.

Jones, K. E., A. Purvis, and J. L. Gittleman. 2003. Biological correlates of extinction risk in bats. *American Naturalist* 161:601–614.

Jonsson, T. and B. Ebenman. 1998. Effects of predator-prey body size ratios on the stability of food chains. *Journal of Theoretical Biology* 193:407–417.

Kant, S. 2000. A dynamic approach to forest regimes in developing economies. *Ecological Economics* 32:287–300.

Kauffman, S.A. 1993. *The Origins of Order: Self-organization and Selection in Evolution.* Oxford, U.K.: Oxford University Press.

Keitt, T. and H. E. Stanley. 1998. Scaling in the dynamics of North American breeding bird populations. *Nature* 393:257–259.

Kent, C. and J. Wong. 1982. An index of littoral zone complexity and its measurement. *Canadian Journal of Fisheries and Aquatic Sciences* 39:847–853.

Kenworthy, L. 1999. Economic integration and convergence: A look at the U.S. states. *Social Science Quarterly* 80:858–869.

Kerr, S. R. 1974. Theory of size distribution in ecological communities. *Journal of the Fisheries Research Board of Canada* 31:1859–1862.

Kimmel, B.I. 1983. Size distribution of planktonic autotrophy and microheterotrophy: Implications for organic carbon flow in reservoir foodwebs. *Archive für Hydrobiologie* 97:303–319.

King, P. B. and H. M. Beikman. 1974. *Geological Map of the United States.* U.S. Geological Survey (USGS) Geological Survey Professional Paper no. 901. Washington, D.C.: USGS.

Kirchner, J.W. and A. Weil. 1998. No fractals in fossil extinction statistics. *Nature* 395:337–338.

Kline, J. D., A. Moses, and R. J. Alig. 2001. Integrating urbanization into landscape-level ecological assessments. *Ecosystems* 4:3–18.

Knapp, S. and J. Mallet. 2003. Refuting refugia? *Science* 300:71–72.

Knutson, M. G., G. Butcher, J. Fitzgerald, and J. Shieldcastle. 2001. *Partners in Flight Bird Conservation Plan for the Upper Great Lakes Plain (Physiographic Area 16).* LaCrosse, Wisc.: U.S. Geological Survey (USGS) Upper Midwest Environmental Sciences Center in cooperation with Partners in Flight.

Kondrashov, A. S. and M.V. Mina. 1986. Sympatric speciation: When is it possible? *Biological Journal of the Linnean Society* 27:201–223.

Korcelli, P. 1977. *An Approach to the Analysis of Functional Urban Regions: A Case Study of Poland.* Report no. RM-77–52. Laxenburg, Austria: International Institute for Applied Systems Analysis.

Kozlowski, J. 1996. Energetic definition of fitness? Yes, but not that one. *American Naturalist* 147:1087–1091.

Krugman, P. R. 1996. *The Self-organizing Economy.* Cambridge, U.K.: Blackwell.

Krummel, J. R., R. H. Gardner, G. Sugihara, R.V. O'Neill, and P. R. Coleman. 1987. Landscape patterns in a disturbed environment. *Oikos* 48:321–324.

Kunin, W. E. and K. J. Gaston. 1993. The biology of rarity: Patterns, causes, and consequences. *Trends in Ecology and Evolution* 8:298–301.

Lambert, D. 2006. Functional convergence of ecosystems: Evidence from body mass distributions of North American late Miocene mammal faunas. *Ecosystems* 9: 97–118.

Lambert, W. D. and C. S. Holling. 1998. Causes of ecosystem transformation at the end of the Pleistocene: Evidence from mammal body mass distributions. *Ecosystems* 1:157–175.

Lawton, J. H. 1986. Surface availability and insect community structure: The effects of architecture and the fractal dimension of plants. Pages 317–331 in B. E. Juniper and T. R. E. Southwood, eds., *Insects and Plant Surfaces*. London: Edward Arnold.

——. 1989. What is the relationship between population density and body size in animals? *Oikos* 55:429–434.

Leaper, R., D. Raffaelli, C. Emes, and B. Manly. 2001. Constraints on body-size distributions: An experimental test of the habitat architecture hypothesis. *Journal of Animal Ecology* 70:248–259.

Legendre, P. and L. Legendre. 2003. *Numerical Ecology*. Amsterdam: Elsevier.

Leopold, A. 1933. The conservation ethic. *Journal of Forestry* 31:636–637.

Lessa, E. P. and R. A. Fariña. 1996. Reassessment of extinction patterns among the late Pleistocene mammals of South America. *Paleontology* 39:651–662.

Levey, D. J. and F. G. Stiles. 1992. Evolutionary precursors of long-distance migration: Resource availability and movement patterns in Neotropical landbirds. *American Naturalist* 140:447–476.

Levin, S. A. 1992. The problem of pattern and scale in ecology. *Ecology* 73:1943–1967.

——. 1998. Ecosystems and the biosphere as complex adaptive systems. *Ecosystems* 1:431–436.

——. 1999. *Fragile Dominion: Complexity and the Commons*. Reading, Mass.: Perseus Books.

Levis, S., J. A. Foley, and D. Pollard. 2000. Large scale vegetation feedbacks on a doubled CO_2 climate. *Journal of Climate* 13:1313–1325.

Lévi-Strauss, C. 1962. *The Savage Mind (Nature of Human Society)*. Chicago: University of Chicago Press.

Lindeman, R. L. 1942. The trophic dynamic aspect of ecology. *Ecology* 23:399–417.

Lipsey, M. W. 1990. *Design Sensitivity: Statistical Power for Experimental Research*. Thousand Oaks, Calif.: Sage.

Loreto, V., L. Pietronero, A. Vespignani, and S. Zapperi. 1995. Renormalization group approach to the critical behavior of the forest-fire model. *Physical Review Letters* 75:465–468.

Lotka, A. J. 1925. *Elements of Physical Biology*. Baltimore: Williams and Wilkins.

Ludwig, D., R. Hilborn, and C. Walters. 1993. Uncertainty, resource exploitation, and conservation: Lessons from history. *Science* 260:17.

Ludwig, D., D. D. Jones, and C. S. Holling. 1978. Qualitative analysis of insect outbreak systems: Spruce budworm and forest. *Journal of Animal Ecology* 47:315–332.

Ludwig, D., B. Walker, and C. S. Holling. 1997. Sustainability, stability, and resilience. *Conservation Ecology* 1 (1): 7. Available at: http://www.consecol.org/vol1/iss1/art7/.

——. 2002. Models and metaphors of sustainability, stability, and resilience. Pages 21–48 in L. H. Gunderson and L. Pritchard Jr., eds., *Resilience and the Behavior of Large-Scale Systems*. Washington, D.C.: Island Press.

Ludwig, J. A. and J. M. Cornelius. 1987. Locating discontinuities along ecological gradients. *Ecology* 68:448–450.

Lundberg, J. and F. Moberg. 2003. Mobile link organisms and ecosystem functioning: Implications for ecosystem resilience and management. *Ecosystems* 6:87–98.

Lynch, K. 1960. *The Image of the City*. Cambridge, Mass.: MIT Press.

Mace, G.M. 1994. An investigation into methods for categorizing the conservation status of species. Pages 293–312 in P.J. Edwards, R.M. May, and N.R. Webb, eds., *Large-Scale Ecology and Conservation Biology*. Oxford, U.K.: Blackwell Science.

MacVicar, T.K. 1983. *Rainfall Averages and Selected Extremes for Central and South Florida*. Technical Publication no. 83-2. West Palm Beach, Fla.: South Florida Water Management District Resource Planning Department.

MacVicar, T.K. and S.S.T. Lin. 1984. Historical rainfall activity in central and southern Florida: Average, return period estimates, and selected extremes. Pages 477–509 in P.J. Gleason, ed., *Environments of South Florida: Present and Past II, Memoir 2*. Coral Gables, Fla.: Miami Geological Society.

Makse, H.A., S. Havlin, and H.E. Stanley. 1995. Modeling urban growth patterns. *Nature* 377:608–612.

Mandelbrot, B. 1983. *The Fractal Geometry of Nature*. New York: W.H. Freeman.

Manly, B.F.J. 1996. Are there clumps in body-size distributions? *Ecology* 77:81–86.

Marquet, P.A. and H. Cofré. 1999. Large temporal and spatial scales in the structure of mammalian assemblages in South America: A macroecological approach. *Oikos* 85:299–309.

Marquet, P.A., M. Fernández, S.A. Navarrete, and C. Valdivinos. 2004. Diversity emerging: Towards a deconstruction of biodiversity patterns. Pages 192–209 in M. Lomolino and L.R. Heaney, eds., *Frontiers of Biogeography: New Directions in the Geography of Nature*. Cambridge, U.K.: Cambridge University Press.

Marquet, P.A., J. Keymer, and H. Cofré. 2003. Breaking the stick in space: Of niche models, metacommunities, and patterns in the relative abundance of species. Pages 64–84 in T.M. Blackburn and K.J. Gaston, eds., *Macroecology: Concepts and Consequences*. Oxford, U.K.: Blackwell Science.

Marquet, P.A., R.A. Quiñones, S. Abades, F. Labra, M. Tognelli, M. Arim, and M. Rivadeneira. 2005. Scaling and power-laws in ecological systems. *Journal of Experimental Biology* 208:1749–1769.

Marshall, J.U. 1989. *The Structure of Urban Systems*. Toronto: University of Toronto Press.

Martin, P.S. 1984. Prehistoric overkill: The global model. Pages 354–403 in P.S. Martin and R. Klein, eds., *Quaternary Extinctions*. Tucson: University of Arizona Press.

Marzluff, J.M. and K.P. Dial. 1991. Life history correlates of taxonomic diversity. *Ecology* 72:428–439.

Matson, P.A., W.J. Parton, A.G. Power, and M.J. Swift. 1997. Agricultural intensification and ecosystem properties. *Science* 277:504–509.

Maturana, H.R. and F.J. Varela. 1987. *The Tree of Knowledge: The Biological Roots of Human Understanding*. Boston: New Science Library.

Maurer, B.A. 1994. *Geographical Population Analysis: Tools for the Analysis of Biodiversity*. London: Blackwell Science.

——. 1998a. The evolution of body size in birds. I. Evidence for non-random diversification. *Evolutionary Ecology* 12:925–934.

——. 1998b. The evolution of body size in birds. II. The role of reproductive power. *Evolutionary Ecology* 12:935–944.

——. 2003. Adaptive diversification of body size: The roles of physical constraint, energetics, and natural selection. Pages 174–191 in T.M. Blackburn and K.J. Gaston, eds., *Macroecology: Concepts and Consequences*. Oxford, U.K.: Blackwell Science.

Maurer, B.A., J.H. Brown, and R.D. Rusler. 1992. The micro and macro in body size evolution. *Evolution* 46:939–953.

Maurer, B.A., E.T. Linder, and D.E. Gammon. 2001. A geographical perspective on the biotic homogenization process: Implications from the macroecology of North American birds. Pages 157–177 in J.L. Lockwood and M.L. McKinney, eds., *Biotic Homogenization*. New York: Kluwer Academic, Plenum.

Maurer, B.A. and M.L. Taper. 2002. Connecting geographical distributions with population processes. *Ecology Letters* 5:223–231.

May, R.M. 1977. Thresholds and breakpoints in ecosystems with a multiplicity of stable states. *Nature* 269:471–477.

——. 1986. The search for patterns in the balance of nature: Advances and retreats. *Ecology* 67:1115–1126.

Mayr, E. 1976. *Evolution and the Diversity of Life*. Cambridge, Mass.: Belknap Press of Harvard University Press.

McGeoch, M.A. and K.J. Gaston. 2002. Occupancy frequency distributions: Patterns, artifacts, and mechanisms. *Biological Reviews* 77:311–331.

Meakin, P. 1993. The growth of rough surfaces and interfaces. *Physical Letters* 235: 189–289.

Meyer, W.B. 1994. Bringing hypsography back in: Altitude and residence in American cities. *Urban Geography* 15:505–513.

Milligan, G.W. 1981. A review of Monte Carlo tests of cluster analysis. *Multivariate Behavioral Research* 16:379–407.

Mills, L.S., M.E. Soulé, and D.F. Doak. 1993. The keystone-species concept in ecology and conservation. *BioScience* 43:219–224.

Milne, B.T. 1988. Measuring the fractal geometry of landscapes. *Applied Mathematics and Computation* 27:67–79.

——. 1997. Applications of fractal geometry in wildlife biology. Pages 32–69 in J.A. Bissonette, ed., *Wildlife and Landscape Ecology*. New York: Springer.

——. 1998. Motivation and benefits of complex systems approaches in ecology. *Ecosystems* 1:449–456.

Milne, B.T., K. Johnston, and R.T.T. Forman. 1989. Scale dependent proximity of wildlife habitat in a spatially-neutral Bayesian model. *Landscape Ecology* 2:101–110.

Milne, B.T., M.G. Turner, J.A. Wiens, and A.R. Johnson. 1992. Interactions between the fractal geometry of landscapes and allometric herbivory. *Theoretical Population Biology* 41:337–353.

Moritz, M.A., J.E. Keeley, E.A. Johnson, and A.A. Schaffner. 2004. Testing a basic assumption of shrubland fire management: How important is fuel age? *Frontiers in Ecology and the Environment* 2:67–72.

Morse, D.R., J.H. Lawton, and M.M. Dodson. 1985. Fractal dimension of vegetation and the distribution of arthropod body lengths. *Nature* 314:731–733.

Morse, D.R., N.E. Stork, and J.H. Lawton. 1988. Species number, species abundance, and body length relationship of arboreal beetles in Bornean lowland rain forest trees. *Ecological Entomology* 13:25–37.

Morton, S.R. 1990. The impact of European settlement on the vertebrate animals of arid Australia: A conceptual model. *Proceedings of the Ecological Society of Australia* 16:201–213.

Moulton, M.P. and J.L. Lockwood. 1992. Morphological dispersion of introduced Hawaiian Finches: Evidence for competition and a Narcissus effect. *Evolutionary Ecology* 6:45–55.

Muradian, R. 2001. Ecological thresholds: A survey. *Ecological Economics* 38:7–24.

Murphy, G.I. 1967. Vital statistics of the Pacific sardine. *Ecology* 48:731–736.

National Geographic Society. 2002. *Field Guide to the Birds of North America.* Washington, D.C.: National Geographic Society.

Newman, M. and R. Palmer. 2003. *Modeling Extinction.* Oxford, U.K.: Oxford University Press.

Nicolis, G. and I. Prigogine. 1977. *Self-organization in Nonequilibrium Systems: From Dissipative Structures to Order Through Fluctuations.* New York: John Wiley and Sons.

Nicolis, J. S. 1986. *Dynamics of Hierarchical Systems: An Evolutionary Approach.* Berlin: Springer.

Niklas, K. J. 1994. *Plant Allometry: The Scaling of Form and Process.* Chicago: University of Chicago Press.

Nittmann J., G. Daccord, and H. E. Stanley. 1985. Fractal growth of viscous fingers: Quantitative characterization of a fluid instability phenomenon. *Nature* 314:141–144.

Noon, B. R. and V. H. Dale. 2002. Broad-scale ecological science and its application. Pages 34–52 in K. J. Gutzwiller, ed., *Applying Landscape Ecology in Biological Conservation.* New York: Springer.

Noy-Meir, I. 1973. Desert ecosystems: Environment and producers. *Annual Review of Ecology and Systematics* 4:355–364.

Nystrom, M. and C. Folke. 2001. Spatial resilience of coral reefs. *Ecosystems* 4:406–417.

Odum, E. P. 1953. *Fundamentals of Ecology.* Philadelphia: W. B. Saunders.

Okuguchi, K. and F. Svidarovsky. 2000. A dynamic model of international fishing. *Keio Economic Studies* 13:471–476.

O'Neill, R. V., D. L. DeAngelis, J. B. Waide, and T. F. H. Allen. 1986. *A Hierarchical Concept of Ecosystems.* Monographs in Population Biology no. 23. Princeton, N.J.: Princeton University Press.

Ostrom, E. 1972. Metropolitan reform: Propositions derived from two traditions. *Social Science Quarterly* 53:474–493.

——. 1990. *Governing the Commons: The Evolution of Institutions for Collective Action.* Cambridge, U.K: Cambridge University Press.

Owen-Smith, R. N. 1998. *Megaherbivores: The Influence of Very Large Body Size on Ecology.* Cambridge, U.K.: Cambridge University Press.

Papageorgiou, G. J. 1980. On sudden urban growth. *Environment and Planning* A 12:1035–1050.

Paulay, G. and C. Meyer. 2002. Diversification in the tropical Pacific: Comparisons between marine and terrestrial systems and the importance of founder speciation. *Integrative and Comparative Biology* 42:922–934.

Perrin, N. 1998. On body size, energy, and fitness. *Functional Ecology* 12:500–502.

Perrings, C., K-G. Mäler, C. Folke, C. S. Holling, and B. Jansson, eds. 1995. *Biodiversity Loss: Economic and Ecological Issues.* Cambridge, U.K.: Cambridge University Press.

Perry, D. A. 1994. Self-organizing systems across scales. *Trends in Ecology and Evolution* 10:241–244.

Peters, R. H. 1983. *The Ecological Implications of Body Size.* Cambridge, U.K.: Cambridge University Press.

Peterson, G. D. 2002a. Contagious disturbance, ecological memory, and the emergence of landscape pattern. *Ecosystems* 5:329–338.

——. 2002b. Estimating resilience across landscapes. *Conservation Ecology* 6 (1): 17. Available at: http://www.consecol.org/vol6/iss1/art17/.

——. 2002c. Forest dynamics in the southeastern United States: Managing multiple stable states. Pages 227–246 in L. H. Gunderson and L. Pichard Jr., eds., *Resilience and the Behavior of Large Scale Ecosystems.* Washington, D.C.: Island Press.

Peterson, G., C.R. Allen, and C.S. Holling. 1998. Ecological resilience, biodiversity, and scale. *Ecosystems* 1:6–18.

Petuch, E.J. 1985. An Eocene asteroid impact in southern Florida and the origin of the Everglades. *Geological Society of America, 98th Annual Meeting* (Orlando, Florida, November 28–31) 17 (7): 688.

——. 1987. The Florida Everglades: A buried pseudoatoll? *Journal of Coastal Research* 9:189–200.

Pickett, S.T.A., W.R. Burch, S.E. Dalton, T.W. Foresman, J.M. Grove, and R. Rowntree. 1997. A conceptual framework for the study of human ecosystems in urban areas. *Urban Ecosystems* 1:185–199.

Pickett, S.T.A., M.L. Cadenasso, J.M. Grove, C.H. Nilon, R.V. Pouyat, W.C. Zipperer, and R. Costanza. 2001. Urban ecological systems: Linking terrestrial, ecological, physical, and socioeconomic components of metropolitan areas. *Annual Review of Ecology and Systematics* 32:127–157.

Pierce, J.L. and A.L. Delbecq. 1977. Organizational structure, individual attitudes, and innovation. *Academy of Management Review* 2:27–37.

Pimm, S.L. 1986. Community stability and structure. Pages 309–329 in M.E. Soulé, ed., *Conservation Biology: The Science of Scarcity and Diversity.* Sunderland, Mass.: Sinauer Associates.

Pimm, S.L. and R.A. Askins. 1995. Forest losses predict bird extinctions in eastern North America. *Proceedings of the National Academy of Sciences, USA* 92:9343–9347.

Plantinga, A.J. and J. Wu. 2003. Co-benefits from carbon sequestration in forests: Evaluating reduction in agricultural externalities from an afforestation policy in Wisconsin. *Land Economics* 79:74–85.

Porter, R.C. 1982. The new approach to wilderness through benefit-cost analysis. *Journal of Environmental Economics and Management* 9:59–80.

Prince, R. and J.B. Rosser Jr. 1985. Some implications of delayed environmental costs for benefit cost analysis: A study of reswitching in the western coal lands. *Growth and Change* 16:18–25.

Printaud, J.C. and T. Jaffré. 2001. Patterns of diversity and endemism in palms in ultramafic rocks in New Caledonia. *South African Journal of Science* 97:535–543.

Pyke, G.H., R. Saillard, and J. Smith. 1995. Abundance of Bristlebirds in relation to habitat and fire history. *Emu* 95:106–110.

Quinn, R.M., K.J. Gaston, and H.R. Arnold. 1996. Relative measures of geographic range size: Empirical comparisons. *Oecologia* 107:179–188.

Raffaelli, D., S. Hall, C. Emes, and B. Manly. 2000. Constraints on body size distributions: An experimental approach using a small-scale system. *Oecologia* 122:389–398.

Rapoport, E.H. 1982. *Aerography: Geographical Strategies of Species.* Oxford, U.K.: Pergamon Press.

Rasmussen, E.M. 1985. El Niño and variations in climate. *American Scientist* 73:168–177.

Rebertus, A.J., G.B. Williamson, and E.B. Moser. 1989. Longleaf pine pyrogenicity and turkey oak mortality in Florida xeric sandhills. *Ecology* 70:60–70.

Restrepo, C., L.M. Renjifo, and P. Marples. 1997. Frugivorous birds in fragmented Neotropical montane forests: Landscape pattern and body mass distribution. Pages. 171–189 in W.F. Laurance and R.O. Bierregaard Jr., eds., *Tropical Forest Remnants: Ecology, Management, and Conservation of Fragmented Landscapes.* Chicago: University of Chicago Press.

Rial, J.A., R.A. Pielke, M. Beniston, M. Claussen, J. Canadell, P. Cox, H. Held, N. De Noblet-Ducoudre, R. Prinn, J.F. Reynolds, and J.D. Salas. 2004. Nonlinearities, feedbacks, and critical thresholds within the Earth's climate system. *Climatic Change* 65:11–38.

Rice, W.R. 1984. Disruptive selection on habitat preference and the evolution of reproductive isolation: A simulation study. *Evolution* 38:1251–1260.

——. 1989. Analyzing tables of statistical tests. *Evolution* 43:223–225.

Rice, W.R. and G. Salt. 1990. The evolution of reproductive isolation as a correlated character under sympatric conditions: Experimental evidence. *Evolution* 44:1140–1152.

Richards, J.F. 1990. Land transformation. Pages 163–178 in B.L. Turner II, W.C. Clark, R.W. Kates, J.F. Richards, J.T. Mathews, and W.B. Meyer, eds., *The Earth as Transformed by Human Action: Global and Regional Changes in the Biosphere over the Past 300 Years*. Cambridge, U.K.: Cambridge University Press.

Rising, J.D. and K.M. Somers. 1989. The measurement of overall body size in birds. *The Auk* 106:666–674.

Robbins, C.S., D. Bystrak, and P.H. Geissler. 1986. *The Breeding Bird Survey: Its First Fifteen Years, 1965–1979*. U.S. Fish and Wildlife Service Resourced Publication no. 157. Washington, D.C.: U.S. Fish and Wildlife Service.

Roeder, K. 1994. A graphical technique for determining the number of components in a mixture of normals. *Journal of the American Statistical Association* 89:487–495.

Ropelewski, C.F. and M.S. Halpert. 1987. Global and regional scale precipitation patterns associated with the El Niño/Southern Oscillation. *Monthly Weather Review* 115:1606–1626.

Rosser, J.B., Jr. 1983. Reswitching as a cusp catastrophe. *Journal of Economic Theory* 31:182–193.

——. 1991. *From Catastrophe to Chaos: A General Theory of Economic Discontinuities*. Boston: Kluwer Academic.

——. 1995. Systemic crises in hierarchical ecological economies. *Land Economics* 71:163–172.

——. 2001. Complex ecologic-economic dynamics and environmental policy. *Ecological Economics* 37:23–37.

——. 2005. Complexities of dynamic forest management policies. Pages 191–206 in S. Kant and R. Albert Berry, eds., *Sustainability, Economics, and Natural Resources: Economics of Sustainable Forest Management*. Dordrecht, Netherlands: Springer.

Rosser, J.B., Jr., C. Folke, F. Günther, H. Isomäki, C. Perrings, and T. Puu. 1994. Discontinuous change in multi-level hierarchical systems. *Systems Research* 11:77–94.

Roughgarden, J. 1989. The structure and assembly of communities. Pages 203–226 in J. Roughgarden, R.M. May, and S.A. Levin, eds., *Perspectives in Ecological Theory*. Princeton, N.J.: Princeton University Press.

——. 1998. *A Primer of Ecological Theory*. Upper Saddle River, N.J.: Prentice-Hall.

Roy, K., D. Jablonski, and K.K. Martien. 2000. Invariant size-frequency distributions along a latitudinal gradient in marine bivalves. *Proceedings of the National Academy of Science* 97:13150–13155.

Ruggiero, A. 2001. Size and shape of the geographical ranges of Andean passerine birds: Spatial patterns in environmental resistance and anisotropy. *Journal of Biogeography* 28:1281–1294.

Ryan, K.C. 2002. Dynamic interactions between forest structure and fire behavior in boreal ecosystems. *Silva Fennica* 36:13–39.

Salt, G.W. 1979. A comment on the use of the term *emergent properties*. *American Naturalist* 113:145–149.

Saphores, J. 2003. Harvesting a renewable resource under uncertainty. *Journal of Economic Dynamics and Control* 28:509–529.

SAS Institute Inc. 1999. *SAS User's Guide: Statistics, Version 5*. Cary, N.C.: SAS Institute.

Schaefer, M.B. 1957. Some considerations of population dynamics and economics in relation to the management of marine fisheries. *Journal of the Fisheries Research Board of Canada* 14:669–681.

Scheffer, M. 1998. *Ecology of Shallow Lakes*. London: Chapman and Hall.

Scheffer, M. and S. Carpenter. 2003. Catastrophic regime shifts in ecosystems: Linking theory to observations. *Trends in Ecology and Evolution* 18:648–656.

Scheffer, M., S. Carpenter, J. Foley, C. Folke, and B. Walker. 2001. Catastrophic shifts in ecosystems. *Nature* 413:591–596.

Scheffer, M., S.H. Hosper, M-L. Meijer, B. Moss, and E. Jeppesen. 1993. Alternative equilibria in shallow lakes. *Trends in Ecology and Evolution* 8:275–279.

Scheffer, M. and E.H. van Nes. 2006. Self-organized similarity: The evolutionary emergence of groups of similar species. *Proceedings of the National Academy of Sciences, USA* 103:6230–6235.

Schindler, D.W. 1977. Evolution of phosphorus limitation in lakes. *Science* 195:260–262.

———. 1990. Experimental perturbations of whole lakes as tests of hypotheses concerning ecosystem structure and function. *Oikos* 57:25–41.

Schindler, D.W., R.H. Hesslein, and M.A. Turner. 1987. Exchange of nutrients between sediments and water after 15 years of experimental eutrophication. *Canadian Journal of Fisheries and Aquatic Sciences* 44:26–33.

Schluter, D. 1996. Ecological causes of adaptive radiation. *American Naturalist* 148:S40–S64.

Schmidt-Nielsen, K. 1984. *Scaling: Why Is Animal Size So Important?* Cambridge, U.K.: Cambridge University Press.

Schoennagel, T., T.T. Veblen, and W.H. Romme. 2004. The interaction of fire, fuels, and climate across rocky mountain forests. *BioScience* 54:661–676.

Schwinghamer, P. 1981. Characteristic size distributions of integral benthic communities. *Canadian Journal of Fisheries and Aquatic Sciences* 38:1255–1263.

Scott, J.A. 1986. *The Butterflies of North America: A Natural History and Field Guide*. Stanford, Calif.: Stanford University Press.

Sendzimir, J. 1998. Patterns of animal size and landscape complexity: Correspondence within and across scales. Ph.D. diss., University of Florida, Gainesville.

Sepkoski, J.J., Jr. 1998. Rates of speciation in the fossil record. *Philosophical Transactions of the Royal Society of London B* 353:315–326.

Sethi, R. and E. Somanathan. 1996. The evolution of social norms in common property resource use. *American Economic Review* 86:766–788.

Shorrocks, B., J. Marsters, I. Ward, and P.J. Evernett. 1991. The fractal dimension of lichens and the distribution of arthropod body lengths. *Functional Ecology* 5:457–460.

Siemann, E. and J.H. Brown. 1999. Gaps in mammalian body size distributions reexamined. *Ecology* 80:2788–2792.

Silva, M. and J.A. Downing. 1995. *CRC Handbook of Mammalian Body Masses*. Boca Raton, Fla.: CRC Press.

Silverman, B.W. 1981. Using kernel density estimates to investigate multimodality. *Journal of the Royal Statistical Society, B* 43:97–99.

——. 1986. *Density Estimation for Statistics and Data Analysis.* Monographs on Statistics and Applied Probability. London: Chapman and Hall.

Simberloff, D. and W. Boecklen. 1981. Santa Rosalia reconsidered: Size ratios and competition. *Evolution* 35:1206–1228.

Simon, H. A. 1955. On a class of skew distributions. *Biometrika* 42:425–440.

——. 1962. The architecture of complexity. *Proceedings of the American Philosophical Society* 106:467–482.

——. 1974. The organization of complex systems. Pages 1–27 in H. H. Pattee, ed., *Hierarchy Theory: The Challenge of Complex Systems.* New York: George Braziller.

Simpson, G. G. 1964. Species density of North American recent mammals. *Systematic Zoology* 13:57–73.

——. 1980. *Splendid Isolation: The Curious History of South American Mammals.* New Haven, Conn.: Yale University Press.

Sinclair, A. R. E., S. Mduma, and J. Brashares. 2003. Patterns of predation in a diverse predator-prey system. *Nature* 425:288–290.

Smith, F. A., S. K. Lyons, S. K. M. Ernest, K. E. Jones, D. M. Kaufman, T. Dayan, P. A. Marquet, J. H. Brown, and J. P. Haskel. 2003. Body mass of late quaternary mammals. *Ecology* 84:3403.

Smith, V. H. 1998. Cultural eutrophication of inland, estuarine, and coastal waters. Pages 7–49 in M. L. Pace and P. M. Groffman, eds., *Successes, Limitations, and Frontiers in Ecosystem Science.* New York: Springer.

Sokal, R. R. and F. J. Rohlf. 1981. *Biometry.* 2d ed. New York: W. H. Freeman.

Solé, R. V. and S. C. Manrubia. 1995. Are rain forests self-organized in a critical state? *Journal of Theoretical Biology* 173:31–40.

——. 1996. Extinction and self-organized criticality in a model of large-scale evolution. *Physical Review* E54:R42–R45.

Soranno, P. A., S. R. Carpenter, and R. C. Lathrop. 1997. Internal phosphorus loading in Lake Mendota: Response to external loads and weather. *Canadian Journal of Fisheries and Aquatic Sciences* 54:1883–1893.

South, J. M. and S. Pruett-Jones. 2000. Patterns of flock size, diet, and vigilance of naturalized Monk Parakeets in Hyde Park, Chicago. *The Condor* 102:848–854.

Spirn, A. W. 1984. *The Granite Garden: Urban Nature and Human Design.* New York: Basic Books.

Sprules, W. G. and M. Munwar. 1986. Plankton size spectra in relation to ecosystem productivity, size, and perturbation. *Canadian Journal of Fisheries and Aquatic Sciences* 43:1789–1794.

Stanley, H. E., L. A. N. Amaral, S. V. Buldyrev, A. L. Goldberger, S. Havlin, H. Leschhorn, P. Maas, H. A. Makse, C-K. Peng, M. A. Salinger, M. H. R. Stanley, and G. M. Viswanathan.1996. Scaling and universality in animate and inanimate systems. *Physica A* 231:20–48.

Stanley, M. H. R., L. A. N. Amaral, S. V. Buldyrev, S. Havlin, H. Leschhorn, P. Maas, M. A. Salinger, and H. E. Stanley. 1996. Scaling behavior in the growth of companies. *Nature* 379:804–806.

Summers, R. and A. Heston. 1991. The Penn World Table (Mark 5): An expanded set of international comparisons, 1950–1988. *Quarterly Journal of Economics* 106:327–368.

Swallow, S. K., P. J. Parks, and D. N. Wear. 1990. Policy-relevant nonconvexities in the production of multiple forest benefits. *Journal of Environmental Economics and Management* 19:264–280.

Terborgh, J. and B. Winter. 1983. A method for siting parks and reserves with special reference to Colombia and Ecuador. *Biological Conservation* 27:45–58.

Thomas, T.M. 1970. *A Detailed Analysis of Climatological and Hydrological Records of South Florida with Reference to Man's Influence upon Ecosystem Evolution*. Coral Gables, Fla.: University of Miami.

Tilman, D. and P. Kareiva. 1997. *Spatial Ecology: The Role of Space in Population Dynamics and Interspecific Interactions*. Princeton, N.J.: Princeton University Press.

Tukey, J.W. 1977. *Exploratory Data Analysis*. Reading, Mass.: Addison-Wesley.

Tuomisto, H. and A. D. Poulsen. 1996. Influence of edaphic specialization on pteridophyte distribution in Neotropical rain forests. *Journal of Biogeography* 23:283–293.

Turelli, M., N.H. Barton, and J.A. Coyne. 2001. Theory and speciation. *TRENDS in Ecology and Evolution* 16:330–343.

Turner, M.G., R.H. Gardner, and R.V. O'Neill. 2001. *Landscape Ecology in Theory and Practice: Pattern and Process*. New York: Springer.

U.S. National Park Service. 1998. *Organ Pipe National Monument—Species List*. Washington, D.C.: U.S. National Park Service.

Vandermeer, J. and S. Maruca. 1998. Indirect effects with a keystone predator: Coexistence and chaos. *Theoretical Population Biology* 54:38–43.

Van der Werff, H. 1992. Substrate preferred Lauraceae and ferns in the Iquitos area, Peru. *Candollea* 47:11–20.

Veit, R.R. and M.A. Lewis. 1996. Dispersal, population growth, and the Allee effect: Dynamics of the House Finch invasion of eastern North America. *American Naturalist* 148:255–274.

Vermeij, G.J. 1991. When biotas meet: Understanding biotic interchange. *Science* 253: 1099–1104.

Vézina, A.F. 1985. Empirical relationships between predator and prey size among terrestrial vertebrate predators. *Oecologia* 67:555–565.

Via, S. 2001. Sympatric speciation in animals: The ugly duckling grows up. *TRENDS in Ecology and Evolution* 16:381–390.

Vigil, J., J.R. Pike, and D. Howell. 2002. *A Tapestry of Time and Terrain*. U.S. Geological Survey (USGS) Geological Investigation Series no. I-2720. Washington, D.C.: USGS.

Vilenkin, B.Y. and V.I. Chikatunov. 1998. Co-occurrence of species with various geographical ranges, and correlation between area size and number of species in geographical scale. *Journal of Biogeography* 25:275–284.

Villard, M.A. and B.A. Maurer. 1996. Geostatistics as a tool for examining hypothesized declines in migratory birds. *Ecology* 77:59–68.

Vitousek, P.M., H.A. Mooney, J. Lubchenco, and J.M. Melilo. 1997. Human domination of Earth's ecosystems. *Science* 277:494–499.

Voss, R.S. and L. Emmons. 1996. Mammalian diversity in Neotropical lowland rain forests: A preliminary assessment. *Bulletin of the American Museum of Natural History* 230:1–115.

Wagener, F.O.O. 2003. Skiba points and heteroclinic bifurcations with applications to the shallow lake system. *Journal of Economic Dynamics and Control* 27:1533–1561.

Walker, B.H., A. Kinsig, and J. Langridge. 1999. Plant attribute diversity and ecosystem function: The nature and significance of dominant and minor species. *Ecosystems* 2:95–113.

Walker, B.H., D. Ludwig, C.S. Holling, and R.M. Peterman. 1981. Stability of semi-arid grazing systems. *Journal of Ecology* 69:473–498.

Walker, B.H. and J.A. Meyers. 2004. Thresholds in ecological and social-ecological systems: A developing database. *Ecology and Society* 9:3.

Walters, C.J. 1986. *Adaptive Management of Renewable Resources*. New York: Macmillan.

Ward, J.H. 1963. Hierarchical grouping to optimize an objective function. *Journal of the American Statistical Association* 58:236–244.

Warner, R.E. 1994. Agricultural land use and grassland habitat in Illinois: Future shock for Midwestern birds? *Conservation Biology* 8:147–156.

Warren, P.H. and J.H. Lawton. 1987. Invertebrate predator-prey body size relationships: An explanation for upper triangular food webs and patterns in food web structure? *Oecologia* 74:231–235.

Watt, A.S. 1947. Pattern and process in the plant community. *Journal of Ecology* 35:1–22.

Webb, S.D. 1985. Late Cenozoic mammal dispersals between the Americas. Pages 357–386 in F. G. Stehli and S. D. Webb, eds., *The Great American Biotic Interchange*. New York: Plenum Press.

———. 1991. Ecogeography and the great American interchange. *Paleobiology* 17:266–280.

Webb, T.J. and K.J. Gaston. 2000. Geographic range size and evolutionary age in birds. *Proceedings of the Royal Society of London* Series B: Biological Sciences 267:1843–1850.

Webster, R. 1978. Optimally partitioning soil transects. *Journal of Soil Science* 29:388–402.

Wen, Y.H.., A. Vezina, and R.H. Peters. 1994. Phosphorus fluxes in limnetic cladocerans: Coupling of allometry and compartmental analysis. *Canadian Journal of Fisheries and Aquatic Sciences* 51:1055–1064.

West, G.B. and J.H. Brown. 2005. The origin of allometric scaling laws in biology from genomes to ecosystems: Towards a quantitative unifying theory of biological structure and organization. *Journal of Experimental Biology* 208:1575–1592.

West, G.B., J.H. Brown, and B.J. Enquist. 1997. A general model for the origin of allometric scaling laws in biology. *Science* 276:122–126.

———. 1999. The fourth dimension of life: Fractal geometry and allometric scaling of organisms. *Science* 284:1677–1679.

Western, D. 2001. Human-modified ecosystems and future evolution. *Proceedings of the National Academy of Sciences USA* 98:5458–5465.

Wiens, J.A. 1984. On understanding a non-equilibrium world: Myth and reality in community patterns and processes. Pages 439–457 in D. Simberloff, L. Aberle, and A.R. Thistle, eds., *Ecological Communities: Conceptual Issues and Evidence*. Princeton, N.J.: Princeton University Press.

———. 1989. Spatial scaling in ecology. *Functional Ecology* 3:385–397.

———. 1995. Landscape mosaics and ecological theory. Pages 1–26 in L. Hansson, L. Fahrig, and G. Merriam, eds., *Mosaic Landscapes and Ecological Processes*. London: Chapman and Hall.

Williams, M. 1990. Forests. Pages 179–201 in B.L. Turner II, W.C. Clark, R.W. Kates, J.F. Richards, J.T. Mathews, and W.B. Meyer, eds., *The Earth as Transformed by Human Action: Global and Regional Changes in the Biosphere over the Past 300 years*. Cambridge, U.K.: Cambridge University Press.

Wilmers, C.C. and W.M. Getz. 2005. Gray wolves as climate change buffers in Yellowstone. *Public Library of Science Biology* 3 (4): e92

Wilson, E.O. 1961. The nature of the taxon cycle in the Melanesian ant fauna. *American Naturalist* 95:169–193.

Wilson, J., B. S. Low, R. Costanza, and E. Ostrom. 1999. Scale misperceptions and the spatial dynamics of a social-ecological system. *Ecological Economics* 31:243–257.
Wilson, M. A. and S. R. Carpenter. 1999. Economic valuation of freshwater ecosystem services in the United States: 1971–1997. *Ecological Applications* 9:772–783.
Woinarski, J. C. Z. 2006. Predictors of nomadism in Australian birds: A re-analysis of Allen and Saunders (2002). *Ecosystems* 9:689–699.
Wu, J. and O. L. Loucks. 1995. From balance of nature to hierarchical patch dynamics: A paradigm shift in ecology. *Quarterly Review of Biology* 70:439–466.
Yule, G. U., A. Stuart, and M. G. Kendall. 1971. *Statistical Papers of George Udny Yule.* New York: Hafner.
Zimov, S. A. 2005. Pleistocene park: Return of the mammoth's ecosystem. *Science* 308:796–798.
Zimov, S. A., V. I. Chuprynin, A. P. Oreshko, F. S.Chapin III, J. F. Reynolds, and M. C. Chapin. 1995. Steppe-tundra transition: A herbivore-driven biome shift at the end of the Pleistocene. *American Naturalist* 146:766–794.
Zinkhan, F. C. 1991. Option pricing and timberland's land-use conversion option. *Land Economics* 67:317–325.
Zipf, G. K. 1949. *Human Behaviour and the Principle of Least Effort: An Introduction to Human Ecology.* Cambridge, Mass.: Addison-Wesley.

CONTRIBUTORS

SEBASTIAN ABADES
Center for Advanced Studies in Ecology and Biodiversity
Departamento de Ecología, Facultad de Ciencias Biológicas
Pontificia Universidad Católica de Chile, Casilla 114-D
Santiago, Chile 6513677

CRAIG R. ALLEN
U.S. Geological Survey, Nebraska Cooperative Fish and Wildlife Research Unit
School of Natural Resources
University of Nebraska
Lincoln, NE, 68585

NATALIA ARANGO
Instituto de Investigación de Recursos Biológicos
Alexander von Humboldt, Carrera 7 #35–20
Bogotá, Colombia 3202767

K. MICHAEL BESSEY
Program in Policy Studies
Clemson University
Clemson, SC 29634

GRAEME S. CUMMING
Percy FitzPatrick Institute of African Ornithology
University of Cape Town
Rondebosch 7701
Cape Town, South Africa

AHJOND S. GARMESTANI
Program in Policy Studies
Clemson University
Clemson, SC 29634

LANCE H. GUNDERSON
Department of Environmental Science
Emory University
Atlanta, GA 30322

TANYA D. HAVLICEK
Center for Limnology
University of Wisconsin
680 North Park Street
Madison, Wisconsin 53706

CRAWFORD S. HOLLING
Department of Zoology
University of Florida
110 Bartram
Gainesville, FL 32611

JUAN E. KEYMER
Department of Ecology and Evolutionary Biology
Princeton University
Princeton, NJ 08544

DONALD LUDWIG
Department of Mathematics
University of British Columbia
Vancouver, BC V6T 1Z2

PABLO A. MARQUET
Center for Advanced Studies in Ecology and Biodiversity
Departamento de Ecología, Facultad de Ciencias Biológicas
Pontificia Universidad Católica de Chile
Casilla 114-D
Santiago, Chile 6513677

BRIAN A. MAURER
Department of Fisheries and Wildlife
Michigan State University
East Lansing, MI 48824

GARRY D. PETERSON
McGill School of the Environment
McGill University
Montreal, Quebec H3A2A7

CARLA RESTREPO
Department of Biology
University of Puerto Rico–Río Piedras
San Juan, PR 00931

J. BARKLEY ROSSER JR.
Program in Economics
James Madison University
Harrisonburg, VA 22807

JAN P. SENDZIMIR
International Institute of Applied Systems Analysis
Laxenburg, Austria A-2361

JENNIFER J. SKILLEN
Department of Fisheries and Wildlife
Michigan State University
East Lansing, MI 48824

CRAIG A. STOW
Duke University
Nicholas School of the Environment
Durham, NC 27708

HORACIO ZEBALLOS
Center for Advanced Studies in Ecology and Biodiversity
Departamento de Ecología, Facultad de Ciencias Biológicas
Pontificia Universidad Católica de Chile
Casilla 114-D
Santiago, Chile 6513677

INDEX